Elliptic Tales

Elliptic Tales

Curves,
Counting,
and
Number Theory

Avner Ash
Robert Gross

Princeton University Press
Princeton and Oxford

Copyright © 2012 by Princeton University Press

Published by Princeton University Press, 41 William Street, Princeton, New Jersey 08540

In the United Kingdom: Princeton University Press, 6 Oxford Street, Woodstock,

Oxfordshire OX20 1TW

press.princeton.edu

Library of Congress Cataloging-in-Publication Data

Ash, Avner, 1949–

Elliptic tales : curves, counting, and number theory / Avner Ash, Robert Gross.

    p. cm.

Includes bibliographical references and index.

ISBN 978-0-691-15119-9 (hardcover)

1. Elliptic functions.   2. Curves, Elliptic.   3. Number theory.

I. Gross, Robert, 1959–   II. Title.

QA343.A97 2012

515′.983—dc23

               2011044712

British Library Cataloging-in-Publication Data is available

This book has been composed in Minion Pro

Typeset by S R Nova Pvt Ltd, Bangalore, India

1 3 5 7 9 10 8 6 4 2

For Edie and Rosemary

Than longen folk to gon on pilgrimages,
And palmeres for to seken strange strondes.
—Geoffrey Chaucer,
The Canterbury Tales

. . . . .

The science of Pure Mathematics, in its modern developments,
may claim to be the most original creation of the human spirit....
The originality of mathematics consists in the fact that in
mathematical science connections between things are exhibited
which, apart from the agency of human reason, are extremely
unobvious. Thus the ideas, now in the minds of contemporary
mathematicians, lie very remote from any notions which can be
immediately derived by perception through the senses.
—Alfred North Whitehead,
Science and the Modern World

Contents

· · ◉ · ·

PART II. ELLIPTIC CURVES AND ALGEBRA

PART III. ELLIPTIC CURVES AND ANALYSIS

Preface

As the pilgrims in Chaucer's *Canterbury Tales* make their way across England to the shrine of St. Thomas Becket in Canterbury, they tell each other all sorts of stories. The aim of *Elliptic Tales* is to tell a variety of mathematical stories, in some depth but at a leisurely pace, relating a few of the beautiful ideas that have been developed to study the solutions of certain cubic equations called "elliptic curves."

The goal of this book is to explain the conjecture of British mathematicians Bryan Birch (1931–) and Peter Swinnerton-Dyer (1927–), made in the 1960s, about the number of solutions to these cubic equations. This is usually referred to as the BSD Conjecture. (If you want to see a statement of the Conjecture right away, turn to p. 236.) In the year 2000, the Clay Mathematics Institute announced a list of seven research problems, which they called "The Millennium Problems," and offered a prize of one million dollars for a solution of each problem. You can read about these problems on their website `http://www.claymath.org/millennium`. The first problem on the list is the BSD Conjecture. A good popular exposition of all seven problems is Devlin (2002); see also Carlson et al. (2006).

Elliptic curves lie in one of the most vibrant areas of mathematics, at the frontiers of research in number theory. An elliptic curve can be described by an equation of the form $y^2 = x^3 + Ax + B$, where A and B are whole numbers. Pretty simple looking, isn't it? But for reasons we explain in this book, equations in two variables like this, where the highest power of the variables is 3, are right at the edge of current research, subject to deeply powerful methods of analysis. We hope that in the end, you will share our enthusiasm for the very beautiful conjecture of Birch and Swinnerton-Dyer.

Our explanations of the ideas and constructions that lie behind the conjecture will involve the telling of many fascinating subsidiary stories of algebra, analysis, and number theory. In some cases, we will develop these topics in more detail than is necessary simply for explaining the BSD Conjecture. We believe that the mathematical ideas revealed in our digressions will be of interest in their own right, just as the pilgrims' stories have an enduring intrinsic value. Indeed, our excursions explain basic topics that are used again and again in many different fields of mathematics.

Elliptic curves belong to the larger class of algebraic curves defined over the rational numbers. Such a curve is defined by a polynomial equation in two variables, with all the coefficients being rational numbers (ordinary fractions). The main number theoretical problem concerning such a curve is to describe, and find if possible, all the rational solutions to this equation. The sorts of questions we ask are:

- Are there any solutions?
- A finite number only?
- If finite, how many solutions are there?
- Infinitely many?
- If infinitely many, how "big" is the solution set?
- Is there any "structure" to the solution set?
- How can one find all the solutions?

This set of questions, especially as applied to elliptic curves, motivate this book. As we go through these questions and explain the concepts that number theorists have developed to ask and answer them, we will be led through several different areas of mathematics.

We have not shied away from details. On some pages, this approach leads to a thicket of algebraic equations. Of course, you may skip any details that seem to be preventing you from seeing the forest for the trees. You can always add the details back later.

Our first book, *Fearless Symmetry* (Ash and Gross, 2006), required of the reader (in theory at least) only a knowledge of algebra. This book also requires some knowledge of basic calculus. We review the bits of calculus that we use, but you do need to know what a derivative is to read some parts of this book.

Besides a Prologue, this book has three parts. Part I is a general exposition of the concept of the degree of an algebraic curve. For example, an elliptic curve has degree 3. As Goldilocks might have said, 3 is neither too big nor too small—it is just right. One point we hope to make is that the theory of elliptic curves is so rich because of the propitious size of the degree.

In Part I, we will explain how the degree of a curve can be geometrically interpreted as the number of points in which a straight line will intersect the curve. We later use this fact when discussing curves of degree 3, but we devote many more chapters to the degree than would be necessary just for this use. The reason is that the logical development of this geometrical interpretation of degree leads us naturally to other important constructions in algebra and algebraic geometry: to the algebraic closure of a field, to the projective plane, and to the concept of multiplicity of a root or an intersection. We motivate each of these constructions from the desire to obtain the same number of intersection points of a line with a given curve no matter which line we use, and this requires us at each stage to enlarge the concept of "intersection point."

Learning the topics in Part I will deepen your understanding of the context for the study of elliptic curves, but it is not strictly necessary to go through this material in such detail. If you wish, you may skip right to Part II and look back at Part I for concepts or definitions as needed. Part II introduces the main characters: elliptic curves. It also surveys a number of algebraic tools for studying them.

Part III is devoted to studying elliptic curves using tools from calculus. It leads finally to an explication of the BSD Conjecture, which unites the algebraic and analytic approaches to the description of elliptic curves.

A certain amount of mathematical sophistication is needed to read this book. We believe that if you've had and enjoyed a college course in calculus or beyond, and if you are patient, you probably have enough of this elusive quality to enjoy any chapter of the book.

In the following table, we tell what mathematical background is needed for the various chapters. There may be small parts of a chapter that need more background than indicated, but they will not be essential parts of the chapter. A knowledge of matrices and determinants might be helpful in a few places, but these ideas are not critical for your enjoyment and understanding of the rest of the book.

Algebra with coordinate geometry: Prologue, Ch. 1–3, 7–10
Algebra with a bit of calculus: Ch. 4–6
A fair amount of calculus: Ch. 11–15

Chaucer's pilgrims were wending their way toward a shrine that memo-
rialized a great event in the past. Our explanations aim at a conjecture,
whose proof (assuming the conjecture is true) awaits us in the future.
Mathematics is always forward-looking in this way. Every great theorem
beckons beyond itself to new mathematical structures and ideas yet to be
discovered. When the BSD Conjecture is proven, glorious as the proof may
be, it will surely become less of a shrine and more of a signpost to yet more
mathematics.

Ways to Read This Book

We ordered the chapters in this book in what seems to us a logical way.
However, many of the chapters are only loosely connected to what comes
before them, so there are several possible orders in which to read them.

1. You could of course read the book straight through. Part I is
   devoted to the algebraic geometry of curves and their
   intersections. Part II develops some tools of an algebraic nature
   for studying elliptic curves. Part III is devoted to describing
   properties of elliptic curves using advanced calculus (which
   mathematicians call "analysis"). It culminates in the explanation
   of the Conjecture of Birch and Swinnerton-Dyer, abbreviated as
   the "BSD Conjecture," about the rank of an elliptic curve.
2. You could read straight through, omitting some things that are
   quite interesting but are not used later in any essential way. You
   could certainly skip all but the first few pages of chapter 5 on
   Bézout's Theorem, the appendix to chapter 7, section 6 of
   chapter 8, the four appendices to chapter 9, which prove certain
   formulas used for singular cubic equations, and the third section
   of chapter 14 on the functional equation of the L-function of an
   elliptic curve.
3. Each of the three parts could be read separately, because each
   part has its own interesting mathematical stories to tell. A later

chapter might use some concepts or propositions from an earlier
chapter, in which case you could look back and fill in the
necessary gap, without having to read all of the previous chapters.

4. If you are eager to get to elliptic curves immediately, you can
begin your reading with Part II and fill in things from earlier
chapters as needed. Even more radically, you could begin with
chapter 15 on the BSD Conjecture, especially if you already know
a bit about elliptic curves.

In all cases, we suggest you read the Prologue for a general orientation
to the subject matter and the Epilogue for a very brief overview of where
things go from here. However you read the book, we hope you enjoy it!

Acknowledgments

. . . . .

First and foremost, we wish to thank Keith Conrad for answering our questions so readily, and especially for his great help in formulating the content of our culminating chapter 15 on the BSD Conjecture.

We extend a gracious thanks to everyone at Princeton University Press who worked on this book, starting with our editor Vickie Kearn, who continually encouraged us, and Anna Pierrehumbert, for many helpful suggestions. Thanks also to our copy editor Charlotte J. Waisner.

A.A. thanks the Clay Mathematics Institute for inviting his participation in the SAGE 18 Days Workshop in December, 2009, which was devoted to computational techniques related to elliptic curves and the BSD Conjecture.

Elliptic Tales

Here we will introduce some of the leading characters in our story. To give some form to our tale, we will consider a particular question: Why were Greek mathematicians able to solve certain kinds of algebraic problems and not others? Of course, the Greeks had no concept corresponding exactly to our "algebra," but we feel free to be anachronistic and to use modern algebraic equations to describe what they were doing. The various concepts and terms we use in the Prologue will be explained in greater detail later in the book.

To start with, let's define an "elliptic curve" as an equation of the form

$$y^2 = x^3 + Ax + B$$

where A and B are fixed integers. For example, $y^2 = x^3 - x$ is an elliptic curve. Here we have set $A = -1$ and $B = 0$. (There is a technical condition we are glossing over here: The cubic polynomial $x^3 + Ax + B$ must have three unequal roots for the equation to define an elliptic curve.) We also use the term "elliptic curve" to denote the set of solutions to this equation.

Be warned: Elliptic curves are *not* ellipses. The terminology stems historically from the fact that in the 1700s, the problem of finding arc-length on an ellipse led, via "elliptic integrals," into the realm of these cubic equations.

How can we solve an equation of this kind, when we want the solutions to be rational numbers (ratios of whole numbers), and how can we describe the set of solutions? How many rational solutions are there? Does the set of solutions admit any special structures or patterns? Can we find all of the rational solutions?

You might think it would be simpler to ask about integral solutions (where x and y are whole numbers), but in fact it is much harder, so we won't do that. The case where x and y may be rational is hard enough, and very beautiful, as you will see.

We call the equation of an elliptic curve a *cubic equation*, because the highest powered term appearing in the equation is the x-cubed term. A synonym for this is to say it is an equation of "degree 3." For another example, the equation $x^2 + y^2 = 1$ is a *quadratic equation*; it is an equation of degree 2. This concept of degree has vital importance for us in this book.

Greek mathematics was primarily geometrical. The first great written work of Greek geometry that has been preserved for us are the thirteen books of Euclid's *Elements*, written about 300 BCE. It is not known exactly how much of the material was discovered by Euclid, or to what extent he invented the orderly presentation in terms of axioms, postulates, theorems, and proofs. In any case, the logical power of Euclid's presentation is enormous and very impressive. If you have studied Euclidean geometry, you know it is largely about figures made out of lines and circular arcs.

Greek mathematics continued to develop after Euclid. Archimedes, who died in 212 BCE during the Roman sack of Syracuse in Sicily, is generally considered one of the greatest mathematicians of all time. Among his many achievements, he dealt brilliantly with the properties of circles, spheres, cylinders, and cones. His methods of proof, like Euclid's, were geometric.

Diophantus, who probably lived around 270 CE in Alexandria, Egypt, was another outstanding Greek mathematician. His work is more akin to algebra than to geometry. His methods do not attain the full scope of our modern algebra, with its symbol-generating abstraction and symbolic manipulation, but he used unknowns combined arithmetically, and solved equations. In fact, the modern field of number theory called "Diophantine equations" is named after him. A Diophantine equation is a polynomial equation in one or more variables whose solutions are sought in the realms of integers or rational numbers. For example, an elliptic curve is a Diophantine equation.

Now, why did Euclid, Archimedes, and other Greek mathematicians have so much success dealing with lines, circles, spheres, circular cylinders, and cones? For more complicated shapes, their results are quite fragmentary. There are some theorems about spirals, conchoids, and other curves, but the body of these results gives the impression of being unsystematic and unsatisfactory, compared with the theorems about lines, circles, and conic sections.

Well, what makes a good contrast with circles, spheres, and cones? A wavy irregular curve or surface is going to be very difficult to study mathematically. (They will have to wait until modern times to enter rigorous mathematics.) Why? It's impossible to describe those shapes exactly using simple equations, or polynomial equations of any degree. So the first thing to note is that circles, spheres, and cones can all be described by simple, polynomial equations. For example, a circle of radius 1 in the xy-plane can be described as the solution set to the equation $x^2 + y^2 = 1$. The sphere of radius 1 in xyz-space can be described by the equation $x^2 + y^2 + z^2 = 1$, and the $45°$ cone can be described by $x^2 + y^2 - z^2 = 0$.

Let's restrict ourselves from now on to shapes that can be described by polynomial equations. These are the equations that Diophantus studied. But even Diophantus was only successful with certain types of problems and not others. His arithmetical investigations were most effective when the powers of the unknowns weren't too big. In fact, the bulk of the problems he dealt with only involved first and second powers. He apparently never even tried problems with any power higher than the sixth. His work includes only a few problems with third, fourth, fifth, and sixth powers.

Diophantus did look at some cubic equations. Archimedes also dealt with some geometric problems that led to cubic equations. These problems were very beautiful but tough. Archimedes solved some of them using conic sections, and the solutions are quite involved. Many of Diophantus's solutions have an *ad hoc* air to them. For some cubical problems, like finding the cube root of 2, Greek mathematicians used various movable mechanical devices, like sliding rulers, that went beyond the usual domain of straightedge and compass. There was something about these cubical problems that took them beyond the usual methods for dealing with lines and circles and which hinted at rich structures that were not understood by the Greeks.

We now know that equations of degree 3 are in some kind of border area between equations of lower degree (easy) and equations of higher degree (very hard). Compared to equations of other degrees, cubic equations enjoy some especially nice, but complex, properties. (Technical point: Certain equations of degree 4 in two variables can be reduced by a certain change of coordinates to equations of degree 3. So when we keep saying "degree 3" we are tacitly including "some degree 4 cases" in our scope. But because of this reduction of 4 to 3, it is economical and fair to talk only about degree 3 equations in our treatment.)

As you will see later in this book, when the degree of an equation in two variables is 3, the solutions of the equation possess some special algebraic structure, called a "group structure." This is a really key point and it has everything to do with the degree being 3. The degree is important because a general line will intersect a curve described by an equation of degree d in d points. So since the degree is 3, a general line will intersect an elliptic curve in exactly 3 points. This observation can be used to create a group structure, which means you can combine two points together to create a third point. The rule is that if P, Q, and R are solutions to the equation of an elliptic curve, and we view them as points on the curve, *à la* Descartes, then $P + Q + R = 0$ if and only if P, Q, and R are the three points of intersection of some line with the elliptic curve. You need exactly three intersection points to make a definition like this. We can then define addition of two points on an elliptic curve by saying $P + Q = -R$ if and only if $P + Q + R = 0$. We will go over all this again very slowly in chapter 8. (If the degree is 4, it may seem that this wouldn't work, since connecting two points on a curve of degree 4 would give a line that intersects the curve in two more points. But as we mentioned, we can sometimes change coordinates to get an equivalent equation of degree 3, and then play the group game.)

It is this group structure that leads to the power and richness of the theory. Another aspect of the theory consists of certain functions, called analytic functions, which are obtained by counting the number of solutions to the equation of an elliptic curve over "finite fields." There is a concept of "rank" of an elliptic curve—in fact, there are *two* concepts of "rank," an *algebraic* rank and an *analytic* rank. The algebraic rank measures "how many" rational solutions there are to the equation of the given elliptic curve, and the analytic rank measures "how many" finite-field solutions there are. We will conclude our journey in this book by explaining a wonderful conjecture, called the "Birch–Swinnerton-Dyer Conjecture," that says that for any given elliptic curve, its algebraic rank always equals its analytic rank. Explaining all these terms and the Birch–Swinnerton-Dyer Conjecture is our ultimate goal.

These ideas, garnered from working on elliptic curves, can be applied to equations of higher degrees, and to systems of several algebraic equations in any number of variables. This is a very active area at the frontiers of research in number theory, and we will say a little bit about it in the Epilogue.

In the remainder of this Prologue, let's see by example why an equation of degree 2 is easier to work with than an equation of degree 3. It is not necessary to go through this on a first reading of the book, if you don't feel up to some algebraic exercises right now.

Our example will be the equation

$$C : x^2 + y^2 = 1.$$

DEFINITION: A solution to an equation is called a *rational solution* if the values that the solution assigns to the variables are all rational numbers.

In Book X, Proposition 29, Lemma 1 of the *Elements*, Euclid proved a formula that can easily be used to find all rational solutions to equation C. In this book, we formulate matters algebraically, and we will solve equation C algebraically (using some geometrical intuition). Euclid formulated and solved his equivalent problem purely geometrically. (In fact, he proved something stronger in that Lemma than what we will do here. We will find all *rational* solutions of the equation, whereas Euclid found all of the *integer* solutions of $x^2 + y^2 = z^2$. It is an exercise in pure algebra, not too difficult, to show that from Euclid's result, you can derive our solution to equation C.)

The equation C has four obvious solutions $x = 1, y = 0; x = -1, y = 0;$ $x = 0, y = 1;$ and $x = 0, y = -1$. Our job is to find all the other solutions.

To explain what we do, it helps to use some geometry, although we could formulate everything using only algebra. We graph the set of all solutions to C, where x and y are real numbers, in the xy-plane, using Cartesian coordinates. Thus each pair (x, y) will be pictured as a point in the plane. Because the solutions to C are exactly the points a unit distance from the origin (by the Pythagorean Theorem), the graph of the solution

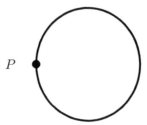

Figure 0.1. The unit circle

set is "the unit circle," that is, the circle C of radius 1 centered at the origin $(0,0)$. You can see a picture of that circle in figure 0.1. In that figure, we single out one particular trivial solution, the point P with coordinates $(-1,0)$, which we will call the "trivial point."

The key fact we are going to use is an algebraic fact: If a straight line L with rational slope hits the circle C at two points Q and R, and if one of those points has rational coordinates, then the other point *has to have* rational coordinates also.

> **DEFINITION**: If a geometric shape K is defined by an equation in several variables, then a point X "on K" is a point whose coordinates are given by a solution to the equation. If this solution is a rational solution—that is, if the coordinates of X are rational numbers—we say that the point X is a *rational point*.

First we will prove the key fact. Then we will show you how to use to it find all rational points on C, starting from the known trivial rational point P.

Suppose the line L is given by the equation $y = mx + b$, where the slope m is a rational number. (In the special case where L is a vertical line, we leave proving the key fact to you as an exercise.) We assume that L hits the circle C at two points Q and R, and that Q has rational coordinates $Q = (s, t)$.

Note that we can now conclude that b is also a rational number. The reason is that s and t are rational and solve the equation of the line L. So $t = ms + b$. Solving for b gives $b = t - ms$, which is a rational number.

Now the points of intersection of L and the circle C can be found by substitution. We are looking for numbers x that solve

$$x^2 + (mx + b)^2 = 1.$$

Multiplying out and collecting terms, this is equivalent to solving

$$(m^2 + 1)x^2 + (2mb)x + (b^2 - 1) = 0.$$

This is a quadratic equation (quadratic! degree 2!) so it factors as

$$(m^2 + 1)(x - \alpha)(x - \beta) = 0$$

where α and β are the two roots. (You could find these roots using the quadratic formula, but we don't need to do that.)

Comparing the last two equations, you see that $\alpha + \beta = (-2mb)/(m^2 + 1)$. Since m and b are rational numbers, so is $\alpha + \beta$. Here is the kicker. We are given by hypothesis that $Q = (s, t)$ is a point on the intersection of L and C and therefore $x = s$ is a solution to the quadratic equation we are studying. In other words, either α or β equals s, which is rational. One of them is rational, and their sum is rational, so the other root is also rational.

Our argument worked only because there were precisely two roots. If we had three roots that summed to a rational number, and we only knew that one of those three roots were rational, there would be no reason to conclude either of the other roots were rational, and in general they wouldn't be.

Going back to our points of intersection Q and R, suppose that $s = \alpha$. Then the other point, R, must have x coordinate equal to β. So $R = (\beta, ?)$. But R is a point on the line L, so $? = m\beta + b$. We see that not only is β rational, but so is $?$. In conclusion, we have proven the following:

KEY FACT: If L is a line with rational slope that meets the circle $C : x^2 + y^2 = 1$ in two points, and if one of those points has rational coordinates, then the other point also has rational coordinates.

OK, how do we use this to find all nontrivial rational points on C? Well, if $Q = (s, t)$ is a rational point in the plane, the line $L = \overline{PQ}$ that connects the point $P = (-1, 0)$ and the point Q has rational slope (since slope = rise/run $= t/(s + 1)$.)

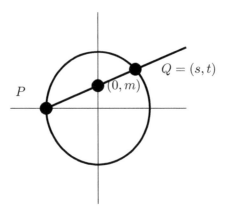

Figure 0.2. The correspondence

Conversely, if L is any line through P with rational slope, it will hit C in exactly one other point, and our key fact means that this new point must have rational coordinates. Our argument shows that there is a one-to-one correspondence between all lines through P with rational slope and all rational points on C.

We have one more computation to mention. The line connecting $(0, m)$ and $P = (-1, 0)$ has slope m. We can now illustrate our one-to-one correspondence in figure 0.2. Every point on the y-axis $(0, m)$ with m rational corresponds to a point (s, t) on the circle with s and t rational, and *vice versa*.

We can now go back to algebra to find the desired formula for all nontrivial rational solutions $x = s$, $y = t$ to C. Each such solution can be viewed as the coordinates of a nontrivial rational point $Q = (s, t)$ on C. Therefore it is the nontrivial point of intersection of the line $L = \overline{PQ}$ and the circle C. The line L has rational slope and so must have an equation of the form

$$y = mx + m$$

with some rational number m, since this line goes through $P = (-1, 0)$.

As we saw above, the x coordinates of P and Q are the solutions to the quadratic equation

$$(m^2 + 1)x^2 + (2mb)x + (b^2 - 1) = 0$$

and these two x-coordinates (which we called α and β in our previous discussion) must add up to $(-2mb)/(m^2 + 1)$. However, now $b = m$, so

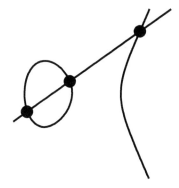

Figure 0.3. $y^2 = x^3 - x$

they add up to $-2m^2/(m^2 + 1)$. Now one of the x-coordinates belongs to P and equals -1. So the other x-coordinate, the one which belongs to Q, is

$$s = -\frac{2m^2}{m^2 + 1} + 1 = \frac{-2m^2 + m^2 + 1}{m^2 + 1} = \frac{1 - m^2}{1 + m^2}.$$

Then

$$t = ms + m = m\frac{1 - m^2}{1 + m^2} + m = \frac{m - m^3 + m^3 + m}{1 + m^2} = \frac{2m}{1 + m^2}.$$

It's always good to check your work. Let's check that in fact $s^2 + t^2 = 1$:

$$s^2 + t^2 = \frac{(1 - m^2)^2}{(1 + m^2)^2} + \frac{(2m)^2}{(1 + m^2)^2} = \frac{1 - 2m^2 + m^4 + 4m^2}{1 + 2m^2 + m^4} = 1.$$

It checks!

Let's look back at a salient feature of what we've done. The key fact was a consequence of the fact that equation C has degree 2. This was crucial. Suppose we picture instead a cubic curve, say, for example

$$E : y^2 = x^3 - x.$$

In figure 0.3, we draw a line with rational slope that hits E in 3 points.

If one of those 3 points has rational coordinates, it is *not* true that either of the other two points of intersection, let alone both, must also have rational coordinates. So drawing lines meeting E does not enable us to produce new rational points from a given one in any obvious way. Things would get even worse with equations of degree higher than 3.

We say that the circle C has a "rational parametrization." This means that the coordinates of every point on the circle are given by a couple of functions that are themselves quotients of polynomials in one variable. The rational parametrization for C that we have found is the following:

$$(x, y) = \left(\frac{1 - m^2}{1 + m^2}, \frac{2m}{1 + m^2} \right).$$

We obtained this rational parametrization by starting with an obvious rational point P, the trivial one, and getting the others by using the secant lines through P. The previous paragraph tells us that this won't work for E and suggests that E does not possess a rational parametrization. This negative result about E can be proven, but giving the proof in this book would take us too far afield.

Our secant method didn't work well with E because we started with just a single rational point. But perhaps we could have started with two rational points. More generally, suppose we have a curve K defined by an equation with rational coefficients in two variables of degree d, with $d > 2$. If a line with rational slope hits K in d points, and if $d - 1$ of those points have rational coordinates, then in fact all d of them have rational coordinates.

Consider now the elliptic curve E (or any other elliptic curve). Because the degree of the equation defining E is exactly 3 and not higher, we can get a new rational point on E if we start from two old ones, at least if the line through the two old ones intersects E in a third point. As mentioned above and explained in detail in chapter 8, this fact allows us to define a binary operation, a kind of addition, on the rational points of E. Since addition takes two things and adds them up to get a third, this technique is going to work well for curves of degree 3 but not in general for those of higher degree. This "group structure" is the starting point of the modern study of cubic equations.

We have seen by example how the degree of an algebraic curve plays a large role in the behavior of the solutions of that curve. Our first task, covered in Part 1, is to try to understand this concept of "degree" in as much detail as possible. This will lead to the developments we outlined in the Preface.

PART I

· · ◈ · ·

DEGREE

Chapter 1

* ⁕ ◉ ◦ ⁕

DEGREE OF A CURVE

Road Map

The idea of *degree* is a fundamental concept, which will take us several chapters to explore in depth. We begin by explaining what an algebraic curve is, and offer two different definitions of the degree of an algebraic curve. Our job in the next few chapters will be to show that these two different definitions, suitably interpreted, agree.

During our journey of discovery, we will often use elliptic curves as typical examples of algebraic curves. Often, we'll use $y^2 = x^3 - x$ or $y^2 = x^3 + 3x$ as our examples.

1. Greek Mathematics

In this chapter, we will begin exploring the concept of the degree of an *algebraic curve*—that is, a curve that can be defined by polynomial equations. We will see that a circle has degree 2. The ancient Greeks also studied lines and planes, which have degree 1. Euclid limited himself to a straightedge and compass, which can create curves only of degrees 1 and 2. A "primer" of these results may be found in the *Elements* (Euclid, 1956). Because 1 and 2 are the lowest degrees, the Greeks were very successful in this part of algebraic geometry. (Of course, they thought only of geometry, not of algebra.)

Greek mathematicians also invented methods that constructed higher degree curves, and even nonalgebraic curves, such as spirals. (The latter cannot be defined using polynomial equations.) They were aware that

Figure 1.1. Three curves

these tools enabled them to go beyond what they could do with straight-edge and compass. In particular, they solved the problems of *doubling the cube* and *trisecting angles*. Both of these are problems of degree 3, the same degree as the elliptic curves that are the main subject of this book. Doubling the cube requires solving the equation $x^3 = 2$, which is clearly degree 3. Trisecting an angle involves finding the intersection of a circle and a hyperbola, which also turns out to be equivalent to solving an equation of degree 3. See Thomas (1980, pp. 256–261, pp. 352–357, and the footnotes) and Heath (1981, pp. 220–270) for details of these constructions. Squaring the circle is beyond any tool that can construct only algebraic curves; the ultimate reason is that π is not the root of any polynomial with integer coefficients.

As in the previous two paragraphs, we will see that the degree is a useful way of arranging algebraic and geometric objects in a hierarchy. Often, the degree coincides with the level of difficulty in understanding them.

2. Degree

We have a feeling that some shapes are simpler than others. For example, a line is simpler than a circle, and a circle is simpler than a cubic curve; see figure 1.1

You might argue as to whether a cubic curve is simpler than a sine wave or not. Once algebra has been developed, we can follow the lead of French mathematician René Descartes (1596–1650), and try writing down algebraic equations whose solution sets yield the curves in which we are interested. For example, the line, circle, and cubic curve in figure 1.1 have equations $x + y = 0$, $x^2 + y^2 = 1$, and $y^2 = x^3 - x - 1$, respectively. On the other hand, as we will see, the sine curve cannot be described by an algebraic equation.

Figure 1.2. $y^2 = x^3 - x$

Our typical curve with degree 3 has the equation $y^2 = x^3 - x$. As we can see in figure 1.2, the graph of this equation has two pieces.

We can extend the concept of equations to higher dimensions also. For example a sphere of radius r can be described by the equation

$$x^2 + y^2 + z^2 = r^2. \tag{1.1}$$

A certain line in 3-dimensional space is described by the pair of simultaneous equations

$$\begin{cases} x + y + z = 5 \\ x - z = 0. \end{cases} \tag{1.2}$$

The "solution set" to a system of simultaneous equations is the set of all ways that we can assign numbers to the variables and make all the equations in the system true at the same time. For example, in the equation of the sphere (which is a "system of simultaneous equations" containing only one equation), the solution set is the set of all triples of the form

$$(x, y, z) = (a, b, \pm\sqrt{r^2 - a^2 - b^2}).$$

This means: To get a single element of the solution set, you pick any two numbers a and b, and you set $x = a$, $y = b$, and $z = \sqrt{r^2 - a^2 - b^2}$ or $z = -\sqrt{r^2 - a^2 - b^2}$. (If you don't want to use complex numbers, and you only want to look at the "real" sphere, then you should make sure that $a^2 + b^2 \leq r^2$.)

Similarly, the solution set to the pair of linear equations in (1.2) can be described as the set of all triples

$$(x, y, z) = (t, 5 - 2t, t),$$

where t can be any number.

As for our prototypical cubic curve $y^2 = x^3 - x$, we see that its solution set includes $(0, 0)$, $(1, 0)$, and $(-1, 0)$, but it is difficult to see what the entire set of solutions is.

In this book, we will consider mostly systems of *algebraic equations*. That means by definition that both sides of the equation have to be polynomial expressions in the variables. The solution sets to such systems are called "algebraic varieties." The study of algebraic geometry, which was initiated by Descartes, is the study of these solution sets. Since we can restrict our attention to solutions that are integers, or rational numbers, if we want to, a large chunk of number theory also falls under the rubric of algebraic geometry.

Some definitions:

- A *polynomial* in one or several variables is an algebraic expression that involves only addition, subtraction, and multiplication of the constants and variables. (Division by a variable is not permitted.) Therefore, each variable might be raised to a positive integral power, but not a negative or a fractional power.
- A *monomial* is a polynomial involving only multiplication, but no addition or subtraction.
- The *degree* of a monomial is the sum of the powers of the variables that occur in the monomial. The degree of the zero monomial is undefined. For example, the monomial $3xy^2z^5$ has degree 8, because $1 + 2 + 5 = 8$.
- The *degree* of a polynomial is the largest degree of any of the monomials in the polynomial. We assume that the polynomial is written without any terms that can be combined or cancelled. For example, the polynomial $3xy^2z^5 + 2x^3z^3 + xyz + 5$ has degree 8, because the other terms have degrees 6, 3, and 0, which are all

smaller than 8. The polynomial $y^5 - x^3 - y^5 + 11xy$ has degree 3, because we first must cancel the two y^5-terms before computing the degree. The polynomial $2x^7y^2 - x^7y^2 + xy - 1 - x^7y^2$ has degree 2, because it is really the polynomial $xy - 1$.

- If we have an algebraic variety defined by a system of equations of the form "some polynomial = some other polynomial," we say that the variety has degree d if the largest degree of any polynomial appearing in the system of equations is d. Again, we assume that the system of equations cannot be simplified into an equivalent system of equations with smaller degree.

EXERCISE: What is the degree of the equation for a sphere in (1.1)? What is the degree of the system of equations for the line in (1.2)?

SOLUTION: The degree of a sphere is 2. The degree of a line is 1.

Now suppose we have a geometric curve or shape. How can we tell what its degree is if we are not given its equation(s)? Or maybe it isn't given by any system of algebraic equations? We're not going to give a general answer to this question in this book, but we will explain the basic idea in the case of single equations.

Let's start by recalling another definition.

DEFINITION: Suppose that $p(x)$ is a polynomial. If b is a number so that $p(b) = 0$, then b is a *root* of the polynomial $p(x)$.

The basic idea is that we will use lines as probes to tell us the degree of a polynomial. This method is based on a very important fact about polynomials: Suppose you have a polynomial

$$f(x) = a_n x^n + a_{n-1} x^{n-1} + \cdots + a_0$$

where $a_n \neq 0$, so that $f(x)$ has degree n, where $n \geq 1$. Suppose that b is a root of $f(x)$. Then if you divide $x - b$ into $f(x)$, it will go in evenly, without remainder.

For example, if $f(x) = x^3 - x - 6$, you can check that $f(2) = 0$. Now divide $x - 2$ into $x^3 - x - 6$ using long division:

$$
\begin{array}{r}
x^2 + 2x + 3 \\
x - 2 \overline{)\, x^3 \qquad\;\; - x - 6} \\
\underline{x^3 - 2x^2 \qquad\qquad} \\
2x^2 - x \\
\underline{2x^2 - 4x} \\
3x - 6 \\
\underline{3x - 6} \\
0
\end{array}
$$

We get a quotient of $x^2 + 2x + 3$, without remainder. Another way to say this is that $x^3 - x - 6 = (x^2 + 2x + 3)(x - 2)$.

We can prove our assertion in general: Suppose you divide $x - b$ into $f(x)$ and get the quotient $q(x)$ with remainder r. The remainder r in polynomial division is always a polynomial of degree less than the divisor. The divisor $x - b$ has degree 1, so the remainder must have degree 0. In other words, r is some *number*. We then take the true statement $f(x) = q(x)(x - b) + r$, and set $x = b$ to get $f(b) = q(b)(b - b) + r$. Since $f(b) = 0$ and $b - b = 0$, we deduce that $r = 0$. So there was no remainder, as we claimed.

Now this little fact has the momentous implication that the number of roots of a polynomial $f(x)$ cannot be greater than its degree. Why not? Suppose $f(x)$ had the roots $b_1, \ldots, b_k$, all different from each other. Then you can keep factoring out the various $x - b_i$'s and get that

$$
f(x) = q(x)(x - b_1)(x - b_2) \cdots (x - b_k)
$$

for some nonzero polynomial $q(x)$. Now multiply out all those factors on the right-hand side. The highest power of x you get must be at least k. Since the highest power of x on the left-hand side is the degree of $f(x)$, we know that k is no greater than the degree of $f(x)$.

Next, we interpret geometrically what it means for b to be a root of the polynomial $f(x)$ of degree n. Look at the graph G of $y = f(x)$. It is a picture in the Cartesian plane of all pairs (x, y) where $y = f(x)$. Now look at the

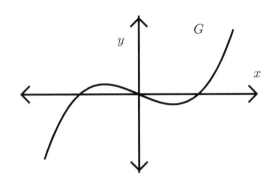

Figure 1.3. $y = x(x - 1)(x + 1)$

Figure 1.4. $y = \sin x$

graph of $y = 0$. It is a horizontal straight line L consisting of all pairs (x, y) where $y = 0$ and x can be anything. OK, now *look at the intersection of the graph of $f(x)$ and the line L*. Which points are in the intersection? They are exactly the points $(b, 0)$ where $0 = f(b)$. That means that the x-coordinates of the points of intersection, the b's, are exactly the roots of $f(x)$. There can be at most n of these roots. This means that the line L can hit the curve G in at most n points. Figure 1.3 contains an example using the function $x^3 - x = x(x - 1)(x + 1)$, which is the right-hand side of our continuing example cubic curve $y^2 = x^3 - x$.

On the other hand, let G be the graph of $y = \sin(x)$. As we can see in figure 1.4, the line L hits G in infinitely many points (namely $(b, 0)$, where b is any integral multiple of π). Therefore, G cannot be the graph of any polynomial function $f(x)$. So the sine wave is not an algebraic curve.

For another example, let H be the graph of $x^2 + y^2 = 1$ (a circle). As we can see in figure 1.5, the line L_1 hits the graph H in 2 points, the tangent line L_2 hits H in one point, and the line L_3 hits H in zero points.

In a case like this, we take the *maximum* number of points of intersection, and call it the *geometric degree* of the curve. So the geometric degree of the circle H is 2. In subsequent chapters, we will discuss how

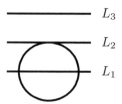

Figure 1.5. Three lines intersecting a circle

mathematicians dealt with the initially unpleasant but ultimately very productive fact that the number of intersection points is not always the same, but depends on the line we choose as probe. The desire to *force* this number to be constant, independent of the probing line, turned out to be a fruitful source of new mathematics, as we will see.

Similarly to the preceding example, we can look at a sphere, for instance the graph of $x^2 + y^2 + z^2 = 1$. Again, a line may intersect the sphere in 2, 1, or 0 points. We say the geometric degree of the sphere is 2.

To repeat, we provisionally *define* the *geometric degree* of a geometric object to be the maximum number of points of intersection of any line with the object. Suppose we have any polynomial $f(x, y, \ldots)$ in any number of variables. It looks very likely from our examples so far that the geometric degree of the graph of $f(x, y, \ldots) = 0$ should equal the degree of f.

3. Parametric Equations

Let's investigate the possibility that the geometric degree of a curve according to the definition of the probing line will equal the degree of the equation of the curve. To do so, we will have to write our probing lines in parametric form.

A *parameter* is an extra variable. A system of equations in parametric form is one where all of the other variables are set equal to functions of the parameter(s). For example, when we describe curves in the xy-plane, we will often use t as a parameter, and write $x = f(t)$ and $y = g(t)$.

We will only use parametric form with a single parameter. A single parameter will suffice to describe curves, and particularly lines. We use parametric form because that method of describing curves makes it easy to find intersection points of two curves by substituting one equation into another, as you will see. Other reasons to use parametric form will also soon become apparent.

Here's an important assumption: *We will always parametrize lines linearly.* What does this assumption mean? Whenever we parametrize a line, we will always set $x = at + b$ and $y = ct + e$, where a and c cannot both be 0. It is possible to pick other more complicated ways to parametrize a line, and we want to rule them out.

We can describe a line or a curve in the plane in two different ways. We can give an equation for it, such as $y = mx + b$. Here, m and b are fixed numbers. You probably remember that m is the slope of the line and b is its y-intercept. A problem with this form is that a vertical line cannot be described this way, because it has "infinite" slope. The equation of a vertical line is $x = c$, for some constant c. We can include all lines in a single formula by using the equation $ax + by = c$ for constants a, b, and c. Depending on what values we assign to a, b, and c, we get the various possible lines.

The other way to describe a line or a curve in the plane is to use a parameter. A good way to think about this is to pretend that the line or curve is being described by a moving point. At each moment of time, say at time t, the moving point is at a particular position in the plane, say at the point $P(t)$. We can write down the coordinates of $P(t) = (x(t), y(t))$. In this way we get two functions of t, namely $x(t)$ and $y(t)$. These two functions describe the line or curve "parametrically," where t is the parameter. The line or curve is the set of all the points $(x(t), y(t))$, as t ranges over a specified set of values.

Parametric descriptions are very natural to physicists. They think of the point moving in time, like a planet moving around the sun. The set of all points successively occupied by the moving point is its "orbit."

For example, the line with equation $y = mx + b$ can be expressed parametrically by the pair of equations $x = t$ and $y = mt + b$. The parametric description may seem redundant: We used two equations where formerly we needed only one. Each kind of representation has its advantages and disadvantages.

One advantage of parametric representation is when we go to higher dimensions. Suppose we have a curve in 6-dimensional space. Then we would need five (or perhaps more) equations in six variables to describe it. But since a line or curve is intrinsically only 1-dimensional, we really should be able to describe it with one independent variable. That's what the parametric representation does for us: We have one variable t and then 6 equations of the form $x_i = f_i(t)$ for each of the coordinates $x_1, \ldots, x_6$ in the 6-dimensional space in which the curve lies.

As we already mentioned, a second advantage of the parametric form is that it gives us a very clear way to investigate the intersection of the line or curve with a geometric object given in terms of equations.

Some curves are easy to express in either form. Consider, for example, the curve with equation $y^2 = x^3$, which can be seen in figure 3.7. This curve can be expressed parametrically with the pair of equations $x = t^2$, $y = t^3$.

On the other hand, the curve defined by the equation $x^4 + 3x^3y + 17y^2 - 5xy - y^7 = 0$ is pretty hard to express parametrically in any explicit way. Conversely, it's hard to find an equation that defines the parametric curve $x = t^5 - t + 1$, $y = e^t + \cos(t)$ as t runs over all real numbers.

For future use, we pause and describe how to parametrize a line in the xy-plane. If a, b, c, and e are any numbers, then the pair of equations

$$\begin{cases} x = at + b \\ y = ct + e \end{cases}$$

parametrizes a line as long as a or c (or both) are nonzero.

Conversely, any line in the xy-plane can be parametrized in this way. In particular: If the line is not vertical, it can be described with an equation of the form $y = mx + b$, and then the pair of equations

$$\begin{cases} x = t \\ y = mt + b \end{cases}$$

parametrizes the same line. If the line is vertical, of the form $x = e$, then the pair of equations

$$\begin{cases} x = e \\ y = t \end{cases}$$

gives a parametrization.

4. Our Two Definitions of Degree Clash

First, let's look at a simple example to show that the degree of an equation doesn't always equal the geometric degree of the curve it defines. In this section, we only consider real, and not complex, numbers. Let K be the graph of $x^2 + y^2 + 1 = 0$. Because squares of real numbers cannot be negative, K is the empty set. Our definition of the probing line would tell us that K would have geometric degree zero. But the polynomial defining K has degree 2.

If you object that we needn't consider empty curves consisting of no points, we can alter this example as follows: Let M be the graph of $x^2 + y^2 = 0$. Now M consists of a single point, the origin: $x = 0$, $y = 0$. By our definition of the probing line, M would have geometric degree 1. But the polynomial defining M has degree 2.

You may still object: M isn't a "curve"—it's got only 1 point. But we can beef up this example: Let N be the graph of $(x - y)(x^2 + y^2) = 0$. Now N consists of the 45°-line, given parametrically by $x = t$, $y = t$. By our definition of the probing line, N would have geometric degree 1. But the polynomial defining N has degree 3.

Another type of example would be the curve defined by $(x - y)^2 = 0$. This is again the 45°-line, but the degree of the equation is 2, not 1.

There is one important observation we can make at this point: If a curve is given by an equation of degree d, *any* probing line will intersect it in *at most d* points. Let's see this by looking at our continuing example. Suppose the curve E is given by the equation $y^2 = x^3 - x$. Take the probing line L given by $x = at + b$, $y = ct + e$ for some real constants a, b, c, and e. As we already mentioned, any line in the plane can be parametrized this way for some choice of a, b, c, and e.

When you substitute the values for x and y given by the parametrization into the equation for E, you will get an equation that has to be satisfied by any parameter value for t corresponding to a point of intersection. If we do the substitution, we get

$$(ct + e)^2 = (at + b)^3 - (at + b).$$

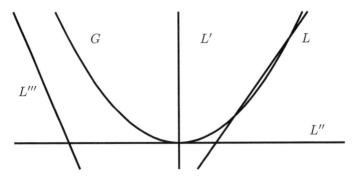

Figure 1.6. Four lines and a parabola

If we expand using the binomial theorem and regroup terms, we obtain the equation

$$a^3t^3 + (3a^2b - c^2)t^2 + (3b^2a - a - 2ec)t + b^3 - b - e^2 = 0$$

We see that the equation will have degree *at most* 3, no matter what a, b, c, and e are. Therefore, it will have at most 3 roots. So at most 3 values of t can yield intersection points of E and L.

Now this lack of definiteness as to the number of intersection points leads to a serious problem with our probing line definition of degree. Let us suppose we have a curve C and we choose a probing line L and we get 5 points in the intersection of C and L. How do we know 5 is the maximum we can get? Maybe a different line L' will yield 6 or more points in the intersection of C and L'. How will we know when to stop probing?

In fact, we can become greedy. We could hope to redefine the geometric degree of C to be the number of points in the intersection of C and L *no matter what line L we pick*! This may sound like a tall order, but if we could do it, we'd have a beautiful definition of degree. It wouldn't matter what probe we choose. Pick any one and count the intersection points.

It may seem like this is hopeless. But let's look at an example. Although it is very simple, this example will show all the problems in our greedy approach to the concept of degree that we will solve in the following three chapters. When we solve them, by redefining the concept of "intersection point" in an algebraically cogent way, our hope will have come true!

Here is the example, which is illustrated in figure 1.6. Let G be the graph of the parabola $y = x^2$. We consider various different probing lines. For example, let L be the line given parametrically by $x = t$, $y = 3t - 2$. The intersection of L and G will occur at points on the line with parameter value t exactly when $3t - 2 = t^2$. This yields the quadratic equation $t^2 - 3t + 2 = 0$, which has the two solutions $t = 1$ and $t = 2$. The two points of the intersection are $(1, 3 \cdot 1 - 2) = (1, 1)$ and $(2, 3 \cdot 2 - 2) = (2, 4)$. This is the optimal case: We get 2 points of intersection, the most possible for an equation of degree 2.

Now look at the probing line L', given parametrically by $x = 0$, $y = t$. This is the vertical line otherwise known as the y-axis. When we plug these values into the equation for the parabola we get $t = 0^2$, which has only one solution: $t = 0$, corresponding to the single point of intersection $(0, 0)$.

The horizontal probing line L'' given by $x = t$, $y = 0$ doesn't fare any better: Plugging in we get $0 = t^2$, which again has only one solution: $t = 0$, corresponding to the single point of intersection $(0, 0)$.

Finally, look at the probing line L''', given parametrically by $x = t$, $y = -5t - 10$. When we plug these values into the equation for the parabola we get $-5t - 10 = t^2$ or equivalently, $t^2 + 5t + 10 = 0$. Using the quadratic formula, we see that since the discriminant $25 - 40 < 0$, there are *no* solutions for t and hence no points of intersection.

Thus, in the simple case of a parabola, we have some probing lines that meet the parabola in 2 points, others that meet the parabola in only 1 point, and others that don't intersect it at all. Yet the equation of a parabola has degree 2. After we finish the next 3 chapters, we will be able to say that *any* probing line intersects the parabola in 2 points, after we have suitably *redefined* the concept of "intersection."

The constructions we will have to make to find a suitable redefinition of "intersection" will be crucial later for our understanding of elliptic curves and so of the Birch–Swinnerton-Dyer Conjecture.

Chapter 2

ALGEBRAIC CLOSURES

Road Map

We saw above that our two definitions of degree might not agree. We would like to *force* them to agree, and the first step in that task is to add elements to our number systems. Adding these "imaginary" numbers will mean that a polynomial equation in a single variable will always have at least one solution.

1. Square Roots of Minus One

The story so far: If we have a polynomial $f(x, y)$ in two variables, we have defined the degree d of f to be the maximum of the degrees of its monomials. If we intersect the graph of $f = 0$ with a parametrized line in the parameter t, that is, a line given by the simultaneous equations $x = at + b$, $y = ct + e$, we get a number N of intersection points. This number N is equal to the number of solutions of the polynomial equation

$$f(at + b, ct + e) = 0,$$

where the lefthand side is a polynomial of degree at most d in the unknown t. PROBLEM: N may not equal d, in which case we find that our desire to define geometrically the degree of an algebraic curve as the number of intersection points of *any* probing line with the curve has been frustrated.

What to do? There is a very interesting book, *Conjectures and Refutations* (Lakatos, 1976) in which the author Imre Lakatos (a Hungarian mathematician who lived from 1922 to 1974) follows an example from topology (Euler's formula for polyhedra) to show how mathematical definitions are not set in stone, but are adapted as the flow of research reveals new discoveries. Apparent contradictions lead to revised definitions removing the contradictions, which lead to new contradictions, and so on. We will see something like this happen with our definition of geometric degree.

Let's look at the worst-case scenario. A curve and a line may not meet at all, even in the plane. For example, a circle of radius r and a line which stays more than r units away from the center of the circle do not have any points of intersection. Nor do parallel lines in a plane. Now the line has degree 1 and the circle has degree 2. We should change our definitions somehow to force that distant line to meet the circle (in 2 points) and to make parallel lines to meet (in 1 point). A third kind of problem occurs when the line is tangent to the circle. Now the line and the circle do meet, but only in 1 point, not 2. We want the intersection in this case also to consist of 2 points, somehow.

These three types of phenomena lead to three extensions of our concept of intersection. The first two problems have in common the fact that they can be solved by extending the domains of our functions. The line and the circle will meet after we add a square root of minus one to our number system, which we discuss in this chapter. The parallel lines will meet by adding "ideal points" to the plane at "infinity." The creation of those ideal points leads to the definition of the "projective plane," which we will discuss in chapter 3. In chapter 4, we will deal with the problem of tangency.

A surprising thing the first two problems have in common is that the minimal amount of extension mentioned in the previous paragraph suffices to solve the corresponding problems with *all* intersections. Just adding a square root of -1 to the real numbers will give us the whole of the complex numbers, which is all we need to make any polynomial in one variable have as many roots as its degree. Similarly, by adding only enough points to make all parallel lines intersect, it will turn out (for example) that we then will have the two intersections we need between the parabola $y = x^2$ and the line described by the y-axis.

The previous paragraph needs to be taken with a grain of salt. A polynomial can have repeated roots, and we have to count the roots with "multiplicity" to ensure that a polynomial of degree d counts as having d roots. This is closely related to the third kind of problem mentioned above, as we will see in chapter 4.

We will follow the Lakatos model in this and the next two chapters. In each chapter, we find that in order to realize our wishes as to the number of intersection points of a line and a curve, we have to extend our concepts of "intersection" and of "points." In this chapter, we will discuss how we enlarge the real numbers to create the algebraically closed field of complex numbers. We will use similar ideas to deal with the finite fields $\mathbf{F}_p$, creating algebraic closures for them as well. Unfortunately, we will need to add more than one new element to each of those fields to get their algebraic closures.

2. Complex Arithmetic

The desire that every nonconstant polynomial with real coefficients have a root leads us to define the complex numbers, denoted by the symbol $\mathbf{C}$. Readers of *Fearless Symmetry*, or those who are conversant with $\mathbf{C}$, can fearlessly skip this section.

We begin by using the symbol i for a square root of -1, in order to have a root of the polynomial $x^2 + 1$. Because no real number has a negative square, i is not a real number. We simply posit its existence, and throw it into the soup of real numbers, where it can combine with them according to the ordinary laws of arithmetic. What is produced is the set of complex numbers, $\mathbf{C}$.

> **DEFINITION**: A *complex number* is a number of the form $a + bi$, where a and b are real numbers.

So $3 + 4i$ and $\sqrt{2} + \pi i$ are complex numbers; so is $0 + 0i$ (which is another way of writing 0). Sometimes, we write the "i" before the second number: $2 - i \sin 2$ is also a complex number. If b is negative, we can write $a - (-b)i$ rather than $a + bi$.

In the complex number $z = a + bi$, we call a the *real part* of z and b the *imaginary part* of z. We write $a = \Re(z)$ and $b = \Im(z)$. We consider every real number x to be a complex number too, but write x instead of $x + 0i$.

Now we need to know how to do arithmetic with complex numbers. Addition and subtraction are simple: We just add or subtract the real and the imaginary parts separately. In other words, $(a + bi) \pm (c + di) = (a \pm c) + (b \pm d)i$.

For multiplication, we use the distributive law of multiplication over addition, which says that $A(B + C) = AB + AC$. Whenever we get $i \cdot i$ (which is also written i^2), we replace it by -1. In other words, we adopt the convention that $i^2 = -1$. Formally, the rule looks like

$$(a + bi)(c + di) = (ac - bd) + (ad + bc)i,$$

because we get the term $(bi)(di) = bdi^2 = -bd$.

Finally, there is division. Rather than tell how to divide one complex number by another, we give a formula for the reciprocal of a complex number. Then, just like with division of fractions, rather than dividing by a complex number, you can multiply by the reciprocal.

First, notice that

$$(a + bi)(a - bi) = a^2 + b^2.$$

We call this quantity the *norm* of the complex number $a + bi$, and write $\mathcal{N}(a + bi) = a^2 + b^2$. Notice that $\mathcal{N}(a + bi)$ is never 0 unless both a and b are 0 (and hence $a + bi = 0$), and that $\mathcal{N}(a + bi)$ is a nonnegative real number. We also define $|a + bi| = \sqrt{a^2 + b^2}$.

DEFINITION: The *complex conjugate* of the complex number $a + bi$ is the complex number $a - bi$.

Here is how we get the reciprocal of the number $a + bi$: We multiply the numerator and denominator by $a - bi$. This gives:

$$\frac{1}{a + bi} = \frac{1}{a + bi} \cdot \frac{a - bi}{a - bi} = \frac{a - bi}{a^2 + b^2} = \frac{a}{a^2 + b^2} - \frac{b}{a^2 + b^2}i. \quad (2.1)$$

So the inverse of $a + bi$ is the rather complicated complex number $\frac{a}{a^2+b^2} - \frac{b}{a^2+b^2}i$. The inverse of $a + bi$ can be written just as $\frac{1}{a+bi}$, but it helps to put all complex numbers into standard $x + yi$ form, so we can compare them. Finally, to divide complex numbers, multiply the dividend by the reciprocal of the divisor.

We will discuss how enlarging **R** to **C** helps us to solve equations, but first we pause for some terminology.

3. Rings and Fields

The words "ring" and "field" might be familiar, but their usage is distinctly different in mathematics. When abstract algebra was developed in the early 1900s, it was found that the same kind of algebraic structures kept recurring in many different areas of mathematics. It was therefore efficient to define each kind of structure and give it a name. Homely names such as "ring" and "field" were usually chosen.

For the next definition, you need to know that an *operation* on a set S is a function that takes two elements of S and gives you back one element of S. For instance, addition is an operation on the set of integers. A subset T of S is *closed* under an operation if whenever the operation is performed on two elements of T, the result is again an element of T.

> **DEFINITION**: A *ring* is a nonempty collection of numbers that is closed under addition, subtraction, and multiplication. Mathematicians often also insist that a ring must contain the number 1, and we will follow that convention.

The definition of ring that you will find in an algebra textbook is more general than the one we have given here. Rings of numbers (hence rings in which multiplication is commutative) will be sufficient for us in this book.

To restate the definition in other words: If R is a ring, then it is a nonempty collection of numbers. The sum, difference, and product of any two of those numbers is again in R, and the number 1 must be in R. Note the glaring absence of "division" in the definition; we will cover that operation in a moment.

Mathematicians have a habit of making definitions as minimal as possible. The number 0 must also be in any ring, but we don't need to put that fact into the definition. Why not? We can pick any element x of the ring R, and then we know that $x - x$ must be in R.

A few examples will convince you that you've been working with rings most of your life.

1. The set of integers **Z** is the most basic example of a ring.
2. The set of rational numbers **Q**, the set of real numbers **R**, and the set of complex numbers **C** are also rings. You can add, subtract, and multiply numbers in these rings without going outside each of them, and each contains the number 1.
3. It is obligatory to give one unfamiliar example. Take the set of all fractions whose denominator, when written in lowest terms, is a power of 2. If n is an integer, we write it as $\frac{n}{1}$, and recall that $1 = 2^0$, so all of the integers are in this collection of numbers. You need to convince yourself now that this collection is closed under addition, subtraction, and multiplication. Notice that it is definitely not closed under division, because 1 and 3 are elements of this ring, while $\frac{1}{3}$ is not an element of this ring.

What role does division play in all of this? We need to be a bit more pedantic here than you might like. If x is an element of a ring R, we don't talk about dividing by x. Instead, we ask whether or not there is an element y in the ring R so that $xy = 1$. You can think of multiplication by this element y as being the same as dividing by x.

DEFINITION: Let x be an element of a ring R. If there is an element y in R so that $xy = 1$, then x is *invertible*, we write $y = x^{-1}$, and we say that y is the *inverse* of x. Notice that it is part of the definition that y be an element of R. If R is any ring, we write $R^\times$ to mean the elements of R which are invertible.

Some more examples:

4. $\mathbf{Z}^\times = \{1, -1\}$.
5. $\mathbf{Q}^\times =$ all fractions other than 0. For if $\frac{a}{b}$ is a fraction, and $a \neq 0$, then $(\frac{a}{b})^{-1} = (\frac{b}{a})$, because $\frac{a}{b} \cdot \frac{b}{a} = \frac{ab}{ba} = 1$.
6. $\mathbf{R}^\times =$ all real numbers other than 0. The fact that you can find a *real* inverse of any nonzero real number must be proven, starting with a definition of **R**. In this book, we will assume that you know that the inversion of a nonzero real number is possible.

7. $\mathbf{C}^\times =$ all complex numbers other than 0. We saw in equation (2.1) how to compute the reciprocal of any nonzero complex number.

8. What about the ring in 3? The invertible elements are $\pm 2^n$, where n is any integer.

DEFINITION: A *field* is a ring with at least two elements in which every element other than 0 is invertible. In other words, if R is a ring, and $R^\times = R - \{0\}$, and $R^\times$ is nonempty, then R is a field.

Again, some examples:

9. $\mathbf{Q}, \mathbf{R},$ and $\mathbf{C}$ are fields.

10. Again, it is obligatory to give an unfamiliar example. This time, we'll take complex numbers of the form $a + bi$, where both a and b are rational numbers. You will need to review the rules for computing $(a + bi)^{-1}$ to convince yourself that if a and b are rational numbers, and $(a + bi)^{-1} = c + di$, then c and d are also rational numbers.

Notice that all fields are rings, but not all rings are fields. For example, $\mathbf{Z}$ and example 3 are not fields.

4. Complex Numbers and Solving Equations

The job of adding on square roots (or cube roots, or anything else that you might like to add on) stops with $\mathbf{C}$. This is the key point. You do not need to add on some other new symbol j with $j^2 = i$, for example; there already is a square root of i in $\mathbf{C}$.

EXERCISE: Let $\alpha = \frac{1}{\sqrt{2}} + \frac{1}{\sqrt{2}}i$. Check that $\alpha^2 = i$.

In fact, every complex number already has a square root in $\mathbf{C}$. But even more than that is true:

THEOREM 2.2: Let $f(x)$ be a nonconstant polynomial whose coefficients are any complex numbers. (For example, $f(x)$ might have integer coefficients.) Then the equation $f(x) = 0$ has solutions in **C**.

Carl Friedrich Gauss, a German mathematician who lived from 1777 to 1855, was the first to prove theorem 2.2. The situation is expressed succinctly by saying that **C** *is an algebraic closure of the field* **R**.

What does "**C** is an algebraic closure of the field **R**" mean? First of all, it means that **R** and **C** are both fields, and that **R** is a subfield of **C** (which means in turn that **R** is a subset of **C** and the results of $+$, $-$, $\times$, and $\div$ in **C** when applied to real numbers give the same answers that they do in **R**). Second of all, it also means that every complex number is the root of some polynomial $f(x)$ with real coefficients. (This is easy to see: $a + bi$ is a solution of $x^2 - 2ax + (a^2 + b^2) = 0$.)

Finally, it means that **C** is itself algebraically closed, which is to say that any polynomial $f(x)$ with complex coefficients has at least one root in **C**. This is the content of theorem 2.2. Note the important fact that theorem 2.2 is equivalent to saying that any nonzero polynomial $f(x)$ with complex coefficients can be factored into linear factors:

$$f(x) = a(x - b_1) \cdots (x - b_n)$$

for some complex numbers $a \neq 0$, $b_1, \ldots, b_n$. If $f(x)$ factors like this, then we see that the degree of f is n, the coefficient of its x^n-term is a, and its roots are $b_1, \ldots, b_n$ (counted with "multiplicity"—see chapter 4).

Let's quickly review the proof of this. Let $f(x)$ be a nonzero polynomial of degree n. If $n = 0$, then $f(x) = a$ is constant. If $n > 0$, then by theorem 2.2, there is some complex root b, that is, some complex number b such that $f(b) = 0$. Now divide $f(x)$ by $x - b$. There will be no remainder: we saw on p. 18 that

$$f(x) = q(x)(x - b). \tag{2.3}$$

for some polynomial $q(x)$ with degree $n - 1$. Now, we can assume by induction that $q(x)$ factors into linear factors, that is, $q(x) = a(x - b_2) \cdots (x - b_n)$. Set $b = b_1$ and plug this factorization into (2.3) to finish the argument.

Let's apply our algebraically closed field **C** to study the intersection of the parabola C defined by the equation $y = x^2$, and the line L, defined by $x = t$, $y = -5t - 10$ (see the end of the last chapter). The intersection points of C and L occur when t satisfies $t^2 + 5t + 10 = 0$. Using the quadratic formula, we find this occurs when

$$t = \frac{-5 + \sqrt{15}i}{2} \quad \text{and} \quad \frac{-5 - \sqrt{15}i}{2}.$$

We thus have two points of intersection, as we hoped, but they are complex points, not real points. These two points are given in coordinates by

$$(x, y) = \left(\frac{-5 + \sqrt{15}i}{2}, -5\frac{-5 + \sqrt{15}i}{2} - 10 \right)$$

and

$$(x, y) = \left(\frac{-5 - \sqrt{15}i}{2}, -5\frac{-5 - \sqrt{15}i}{2} - 10 \right).$$

If we could see into the complex plane with our imagination, we would see C and L intersecting in the complex plane at these two points.

5. Congruences

In addition to the rings and fields in the examples above, we will need some others in order to explain the material at the heart of this book. Congruences between integers are a key to modern number theory. Congruences modulo a positive integer $n > 1$ are best studied in terms of a ring denoted $\mathbf{Z}/n\mathbf{Z}$. It is a finite ring containing n elements. Describing it is our goal in this section. If p is a prime, then $\mathbf{Z}/p\mathbf{Z}$ is not just a ring, but a field, and these fields with a prime number of elements are among the most useful fields that mathematicians have available in their toolkit. Readers of *Fearless Symmetry* can again fearlessly skip this section, or skim it to refresh their memories.

We define two numbers as *congruent modulo n* if they have the same remainder upon division by a positive integer n. For example, we will call 23 and 45 congruent modulo 11, because both 23 and 45 leave the remainder of 1 when divided by 11. The notation invented by Gauss to express this is

$$23 \equiv 45 \quad (\text{mod } 11). \tag{2.4}$$

DEFINITION: We write $a \equiv b$ (mod n) (read as "a is congruent to b modulo n") if a and b have the same remainder when divided by n. Equivalently, we write $a \equiv b$ (mod n) if $a - b$ is a multiple of n. The number n here is the *modulus* of the congruence.

The $\equiv$ symbol can be used in most of the ways that the $=$ symbol is used. If $a \equiv b$ (mod n) and $b \equiv c$ (mod n), then $a \equiv c$ (mod n); this is the analogue for congruences of Euclid's famous axiom, "Things equal to the same thing are equal to each other."

Here is another numerical example to help you understand the notation: $100 \equiv 23$ (mod 11), because both leave a remainder of 1 when divided by 11. From this congruence and the congruence (2.4), we can conclude that $100 \equiv 45$ (mod 11) without having to check again that both 100 and 45 leave the same remainder when divided by 11.

But we can also deal with unknowns. For instance, if $x \equiv y$ (mod 11), then we can see that $x + 2 \equiv y + 13$ (mod 11) and this congruence remains true whatever x and y are. (It is because $2 \equiv 13$ (mod 11); this is like adding equals to equals.) However, we cannot actually divide either side of the congruence by 11 to see what the remainders are, because we do not know what x and y actually are.

Negative numbers can be included in the game. We will always use a positive remainder (or 0), even if we are dividing into a negative number. So, for example, -23 has a remainder of 10 when divided by 11, because 11 goes into -23 just -3 times, leaving a remainder of 10. In an equation, $-23 = -3 \cdot 11 + 10$.

What are all the positive numbers satisfying $x \equiv 5$ (mod 7)? They are 5, 12, 19, 26, 33, 40, . . . (just keep adding 7). They make up what is called an "arithmetic progression." It may have occurred to you that we left negative numbers out of the arithmetic progression. If so, you are right. Besides 5, 12, 19, ..., we should include $-2, -9, -16, -23, ...$ (just keep subtracting 7) in the list of *all* numbers congruent to 5 modulo 7.

We can also use congruence notation with multiplication (we will get to division later). It takes a bit of algebra to show that if $a \equiv b$ (mod 12) and $c \equiv d$ (mod 12), then $a + c \equiv b + d$ (mod 12) and $ac \equiv bd$ (mod 12). For example, $3 \equiv 15$ (mod 12) and $8 \equiv 80$ (mod 12), so $3 + 8 \equiv 15 + 80$ (mod 12), and $3 \cdot 8 \equiv 15 \cdot 80$ (mod 12). In other words, $11 \equiv 95$ (mod 12) and $24 \equiv 1200$ (mod 12).

> **CAUTION**: We can only add, subtract, and multiply congruences if they have the *same* modulus: We have $2 \equiv 7$ (mod 5) and $3 \equiv 13$ (mod 5), and so $2 \cdot 3 \equiv 7 \cdot 13$ (mod 5). But $2 \equiv 7$ (mod 5) and $3 \equiv 13$ (mod 10) do not imply that $2 \cdot 3 \equiv 7 \cdot 13$ (mod 10), which is not true.

> **EXERCISE**: Show that if $a \equiv b$ (mod n) and $c \equiv d$ (mod n), then $ac \equiv bd$ (mod n).

> **SOLUTION**: The hypotheses tell us that $a - b$ and $c - d$ are multiples of n. Therefore $ac - bd = a(c - d) + d(a - b)$ is also a multiple of n.

If n is any integer larger than 1, we use the notation $\mathbf{Z}/n\mathbf{Z}$ to refer to the collection of integers where our computations are done modulo n rather than using ordinary integers. We can think of $\mathbf{Z}/n\mathbf{Z}$ as the collection of integers $\{0, 1, ..., n - 1\}$. All of the facts above add up to the conclusion that $\mathbf{Z}/n\mathbf{Z}$ is a ring.

What about division? Or, to use our new terminology, what about multiplicative inverses? Here is where we have to be careful. In $\mathbf{Z}/10\mathbf{Z}$, for example, we can compute that $1^{-1} = 1$, $3^{-1} = 7$, $7^{-1} = 3$, and $9^{-1} = 9$. The remaining 6 elements—0, 2, 4, 5, 6, and 8—are not invertible.

We have all of these noninvertible elements because 10 is a composite number. In fact, something similar always happens whenever the modulus is a composite number.

EXERCISE: Suppose that n is a composite number. Show that $\mathbf{Z}/n\mathbf{Z}$ is not a field.

SOLUTION: If n is a composite number, we know by definition that we can write $n = ab$, where both a and b are larger than 1, and neither a nor b are as large as n. We claim that a cannot be invertible.

Why? Suppose that we can find a c so that $ac \equiv 1$ (mod n). Multiply through by b, and we get the congruence $abc \equiv b$ (mod n). But $ab = n$, which means that $ab \equiv 0$ (mod n), and multiplying by c gives the congruence $abc \equiv 0$ (mod n). We conclude that $b \equiv 0$ (mod n). That's a contradiction, because $0 < b < n$.

You might guess (correctly) from the listing of elements in $(\mathbf{Z}/10\mathbf{Z})^{\times}$ above that the following is true:

THEOREM 2.5: Let n be an integer larger than 1. The elements of $(\mathbf{Z}/n\mathbf{Z})^{\times}$ consist of those numbers between 1 and $n-1$, that are relatively prime[1] to n.

Recall that a prime number is an integer larger than 1 with no positive divisors other than itself and 1. Theorem 2.5 makes it easy to verify:

THEOREM 2.6: If p is a prime, then $\mathbf{Z}/p\mathbf{Z}$ is a field.

PROOF: Let p be a prime. We need to show that $(\mathbf{Z}/p\mathbf{Z})^{\times}$ consists of all of the numbers between 1 and $p-1$. Theorem 2.5 tells us that $(\mathbf{Z}/p\mathbf{Z})^{\times}$ consists of those numbers between 1 and $p-1$, that are relatively prime to p. Because p is a prime, *every* number between 1 and $p-1$ is relatively prime to p, so $(\mathbf{Z}/p\mathbf{Z})^{\times} = \{1, 2, \ldots, p-1\} = \mathbf{Z}/p\mathbf{Z} - \{0\}$. By definition, that shows that $\mathbf{Z}/p\mathbf{Z}$ is a field.

---

[1] We say that the integers a and b are *relatively prime* if their only common positive factor is 1. A proof of theorem 2.5 may be found in any book about elementary number theory.

Table 2.1. Addition and multiplication in $\mathbf{F}_5$

+	0	1	2	3	4
0	0	1	2	3	4
1	1	2	3	4	0
2	2	3	4	0	1
3	3	4	0	1	2
4	4	0	1	2	3

×	0	1	2	3	4
0	0	0	0	0	0
1	0	1	2	3	4
2	0	2	4	1	3
3	0	3	1	4	2
4	0	4	3	2	1

6. Arithmetic Modulo a Prime

There is a special symbol used when working with the integers modulo p (where p is any prime): $\mathbf{F}_p$. The letter "F" here stands for *field*. So $\mathbf{Z}/p\mathbf{Z}$ and $\mathbf{F}_p$ are the same thing.

$\mathbf{F}_p$ is defined as the set $\{0, 1, \ldots, p-1\}$ with addition and multiplication defined as follows: If x, y, and z are in $\mathbf{F}_p$, $x + y = z$ in $\mathbf{F}_p$ exactly when $x + y \equiv z \pmod{p}$ and $xy = z$ in $\mathbf{F}_p$ exactly when $xy \equiv z \pmod{p}$. In other words, we allow ourselves to use equality signs in $\mathbf{F}_p$ where we would use congruence signs among integers.

We say that $\mathbf{F}_p$ is a "field with p elements." The number system $\mathbf{F}_p$ has *characteristic p*. (A field has characteristic p if $\underbrace{a + a + \cdots + a}_{p \text{ times}} = 0$ for any element a in the field.) We can write out the addition and multiplication tables for any specific prime p.

EXERCISE: Write out the addition and multiplication tables for $\mathbf{F}_5$. Remember that every integer outside of the range 0–4 must be replaced by its remainder modulo 5.

SOLUTION: The answer is in table 2.1. For example, to explain the entry in the second table for the spot in the row labeled "3" and the column labeled "4," we multiply $3 \cdot 4$, getting 12, and divide by 5, getting a remainder of 2. Then we know that $3 \cdot 4 \equiv 2 \pmod{5}$, so we put a 2 in the table.

7. Algebraic Closure

We explained above that theorem 2.2 says that $\mathbf{C}$ is the algebraic closure of $\mathbf{R}$. In fact, if F is *any* field, there is another field, F^{ac}, which is *its* algebraic closure. If $F = F^{\text{ac}}$, then we say that F is *algebraically closed*. For example, $\mathbf{C}$ is algebraically closed.

That's too abstract for us, so let's concentrate on $\mathbf{F}_p$. We will explain why for every prime p, the field $\mathbf{F}_p$ has an algebraic closure, which we will denote by $\mathbf{F}_p^{\text{ac}}$. This is a bigger field that contains $\mathbf{F}_p$ as a subfield. The algebraic closure of $\mathbf{F}_p$ is not unique. (But neither is $\mathbf{C}$ unique! It depends on choosing a square root of -1. If we choose i and you choose j to be the symbol for a square root of -1, we will get different versions of the complex numbers. Of course, they will be "isomorphic," i.e., essentially the same. This is true for $\mathbf{F}_p^{\text{ac}}$ also: No matter how you construct it, you will get something isomorphic to our version of it.)

By definition, $\mathbf{F}_p^{\text{ac}}$ is a big field (it will have an infinite number of elements) containing $\mathbf{F}_p$ as a subfield. Every element of $\mathbf{F}_p^{\text{ac}}$ will be the root of some polynomial with $\mathbf{F}_p$ coefficients. Finally, every polynomial $f(x)$ with coefficients in $\mathbf{F}_p^{\text{ac}}$ will have a root in $\mathbf{F}_p^{\text{ac}}$, so that $f(x)$ can be factored into linear factors with coefficients in $\mathbf{F}_p^{\text{ac}}$, just as we saw above with complex polynomials. (The proof of this fact about polynomial factorization is the same argument as the one we gave above for $\mathbf{C}$.)

THEOREM 2.7: An algebraic closure $\mathbf{F}_p^{\text{ac}}$ of $\mathbf{F}_p$ exists.

Rather than prove theorem 2.7, let's look at an example, to get a flavor of what is going on. To keep things smallish, let's look at the smallest prime, and set $p = 2$. Now $\mathbf{F}_2 = \{0, 1\}$. Perhaps it is already algebraically closed? After all, in this field, we *do* have a square root of -1. Modulo 2, $-1 = 1$ and $1^2 = 1$. But there is another quadratic polynomial, $f(x) = x^2 + x + 1$, and $x^2 + x + 1$ does not have a root in $\mathbf{F}_2$. If you plug in $x = 0$, you get $f(0) = 0 + 0 + 1 = 1 \neq 0$, and if you plug in $x = 1$, you get $f(1) = 1 + 1 + 1 = 1 \neq 0$.

Remember that the characteristic of $\mathbf{F}_2$ is 2, so that $a + a = 0$ for any a in $\mathbf{F}_2$. (More generally, the characteristic of $\mathbf{F}_p$ is p. In fact the distributive law of multiplication implies that $\mathbf{F}_p^{\text{ac}}$ also has characteristic p. For any number a in $\mathbf{F}_p^{\text{ac}}$, $\underbrace{a + a + \cdots + a}_{p \text{ times}} = (1 + 1 + \cdots + 1)a = 0 \cdot a = 0$, where we knew that $1 + 1 + \cdots + 1 = 0$ because 1 is in $\mathbf{F}_p$.)

OK, if we hope to construct an algebraic closure of $\mathbf{F}_2$, we will have to throw a root of $x^2 + x + 1 = 0$ into the soup. Let's posit the existence

Table 2.2. Addition and multiplication in $\mathbf{F}_4$

+	0	1	α	$1+\alpha$	$\times$	0	1	α	$1+\alpha$
0	0	1	α	$1+\alpha$	0	0	0	0	0
1	1	0	$1+\alpha$	α	1	0	1	α	$1+\alpha$
α	α	$1+\alpha$	0	1	α	0	α	$1+\alpha$	1
$1+\alpha$	$1+\alpha$	α	1	0	$1+\alpha$	0	$1+\alpha$	1	α

of a root, and call the root α. This is analogous to how we posited i to be a square root of -1 when we formed $\mathbf{C}$, so we get a kind of complex numbers modulo 2, consisting of all "numbers" of the form $a + b\alpha$ with a and b in $\mathbf{F}_2$. We get a set called $\mathbf{F}_4$, containing 4 elements: 0, 1, α, and $1 + \alpha$. One can check that $\mathbf{F}_4$ is a field, if you use common algebra to add and multiply. The addition and multiplication tables are in table 2.2. You can see that dividing by α is the same as multiplying by $1 + \alpha$, in the same way that when we work with complex numbers, dividing by $a + bi$ is the same as multiplying by $\frac{a-bi}{a^2+b^2}$.

Unfortunately, $\mathbf{F}_4$ is not algebraically closed either. Unlike the case of $\mathbf{C}$, where adjoining a single new "number," namely i, was enough to complete our game, in the case of $\mathbf{F}_2$ it will not be enough merely to adjoin α.

EXERCISE: Find a polynomial with coefficients in $\mathbf{F}_4$ that doesn't have a root in $\mathbf{F}_4$.

SOLUTION: One possibility is $x^4 - x - 1$, because we can see from the multiplication table for $\mathbf{F}_4$ that every element in $\mathbf{F}_4$ satisfies $x^4 - x = 0$.

So we have to adjoin a root, call it β, of the polynomial $x^4 - x - 1$. We'll get a bigger field containing elements like $1 + \alpha + \beta$. In fact, if K is any field of characteristic 2 with a finite number of elements, we can always find a polynomial that does not have a root in K. But there is a clever way to figure out all the roots we will need and add them to $\mathbf{F}_2$ all at once, and then we will have constructed the infinite field $\mathbf{F}_2^{\mathrm{ac}}$.

In fact, it turns out that for any prime p, if we posit and adjoin to $\mathbf{F}_p$ all the roots of all polynomials of the form $x^n - 1$ for all positive n not divisible by p, and keep careful track of what's going on, we will be able to construct $\mathbf{F}_p^{ac}$. For the rest of this book, we only need to know of the existence of $\mathbf{F}_p^{ac}$, so we won't go further here. But if you want to do advanced computations with elliptic curves, or other algebraic varieties defined with integer coefficients, you will have to learn how to work in $\mathbf{F}_p^{ac}$ in detail.

Notice that in constructing $\mathbf{F}_p^{ac}$, we had to construct lots of finite fields. The following two theorems summarize the basic facts about finite fields:

THEOREM 2.8: Suppose that F is a field with a finite number of elements m. Then m must be a power of a prime, that is, $m = p^e$ for some prime p and some integer $e \geq 1$.

THEOREM 2.9: Suppose that $\mathbf{F}_p^{ac}$ is an algebraic closure of $\mathbf{F}_p$. Then $\mathbf{F}_p^{ac}$ contains exactly one subfield with p^e elements for each integer $e \geq 1$.

. . . . .

Chapter 3

THE PROJECTIVE PLANE

Road Map

Adding elements to our number systems is not sufficient to make the two definitions of degree agree. Next, we must add elements to the plane, changing the *affine plane* to the *projective plane*, to force there to be more intersection points between a line and a curve.

1. Points at Infinity

The story so far: If we have a polynomial $f(x, y)$ in two variables of degree d, and if we intersect the graph of $f = 0$ with a parametrized line given by the simultaneous equations $x = at + b$, $y = ct + e$, we get a number N of intersection points. This number N is equal to the number of solutions of the polynomial equation

$$f(at + b, ct + e) = 0,$$

where f is a polynomial of degree at most d. PROBLEM: N may not equal d, in which case we experience the frustration of our desire to define geometrically the degree of an algebraic curve as the number of intersection points of *any* probing line with the curve.

In the previous chapter, we started to solve the problem, by defining the complex numbers. If we allow complex coordinates for our points, then we have more hope of finding d solutions to $f(at + b, ct + e) = 0$ if f has degree d. Two problems remain. One problem is that the polynomial $f(at + b, ct + e)$ could have repeated roots. We will deal with that in the next chapter. The other problem is that $f(at + b, ct + e)$ could have

Figure 3.1. Two intersecting lines

degree smaller than d even though f itself has degree d. We will now tackle this problem.

Let's begin with the simplest possible example of the problem. Suppose we have two lines L and M, where L is given by the equation $ax + by = c$, and M by the equation $dx + ey = f$. Since we have enlarged our number system to include all the complex numbers, we should let a, b, c, d, e, and f be any complex numbers. However, to be able to draw pictures, we shall assume they are real numbers. Also, for L to be a line, at least one of a and b should be nonzero, and the same for M: at least one of d and e must be nonzero.

Finding the intersection point of L and M is algebraically equivalent to solving the simultaneous equations

$$\begin{cases} ax + by = c \\ dx + ey = f. \end{cases}$$

Using any method you like to do this, you will find that the solution exists and is unique if $ae - bd \neq 0$. That inequality is the "generic" case, and we are happy with the result: The lines L and M intersect in the one point whose coordinates we obtain by solving the simultaneous equations. The result is in figure 3.1.

For example, let's solve

$$\begin{cases} 3x + 4y = 10 \\ 2x - 3y = 1. \end{cases}$$

Multiply the first equation by 2 and the second by 3 to get the equivalent system

$$\begin{cases} 6x + 8y = 20 \\ 6x - 9y = 3. \end{cases}$$

Subtract the lower equation from the upper one to get $17y = 17$, from which we conclude that $y = 1$. Substituting this result back into the first equation gives $3x + 4 = 10$, from which we conclude that $x = 2$. So the solution is $(x, y) = (2, 1)$.

It is instructive to see what goes wrong if $ae = bd$. Try to solve

$$\begin{cases} 3x + 4y = 10 \\ 2x + \frac{8}{3}y = 7. \end{cases}$$

The same procedure yields

$$\begin{cases} 6x + 8y = 20 \\ 6x + 8y = 21, \end{cases}$$

and this system of equations has no solutions. If we obstinately continue the same procedure, we obtain $0y = -1$, which leads to the nonsensical equation $y = -\frac{1}{0}$.

Geometrically, if $ae = bd$, then L and M either coincide and are the same line, or else they are parallel, different lines. In the first case, it is silly to ask for L and M to intersect in just one point. There is no way to sneak out of the fact that L and M are the same line, and intersect in all their points.[1]

Let us assume that $ae - bd = 0$ but that L and M are different lines. In this case, L and M are parallel lines. They do not intersect, but we want them to intersect. In a fashion reminiscent of how we simply posited a square root of -1, we shall posit an intersection point for L and M. We call that intersection point "the point at infinity" on both L and M. We often say that railroad tracks appear to meet "at infinity," which is a similar idea. As another example, we say that the vanishing point in one-point perspective "lies at infinity."

---

[1] In general, if two curves "share a component," we will not ask about the number of their points of intersection. If one curve C is given by the equation $f(x, y) = 0$ and the other curve D by the equation $g(x, y) = 0$, then we say that the curves *share a component* if and only if there is some nonconstant polynomial $h(x, y)$ such that f and g are both multiples of h. That is, you can divide h into f and g evenly without leaving any remainders. When this happens, the curve defined by the equation $h(x, y) = 0$ will be part of both C and D. We call it a "common component" of C and D (or a union of common components, if h itself can be factored in a nontrivial way). We will not consider the case when two curves have a common component any further in this book, except to forbid it when we state certain theorems.

If we take a third line N parallel to both L and M, then it too should intersect with each of L and M. It turns out that thrift is the best policy here: We can reuse the point from the previous paragraph. So all three lines "meet" in the same point "at infinity." If you look down subway tracks with a parallel third rail, *all three* rails will appear to meet at a single point "at infinity," and similarly with one-point perspective.

The rule of parsimony thus leads us to add just one point to the plane, through which go all the lines parallel to L and M. To give this point an unambiguous name, we could call it P_Π, where Π is the set of all lines parallel to L and M. The advantage of this notation is that we don't single out a special line or lines, but refer to a set of *all* lines parallel to each other. However, later in this chapter we will develop an even better way of naming these points.

This terminology "points at infinity" comes about because we think of all the points in the xy-plane that we can designate with coordinates as "finite" points (because their coordinates are finite numbers). In some sense, the point at infinity has coordinates one or both of which is infinite. For example, suppose L is the line given by $x = 1$ and M is the line given by $x = 2$. Points on L have coordinates $(1, t)$ and points on M have coordinates $(2, t)$, where t is any real number. We want to say that L and M meet at the point $(1, \infty) = (2, \infty)$. Why? Well, ∞ is so big that it swamps out the finite difference between the 1 and the 2 in the x-coordinate. If you were standing far out there, where the y-coordinate equals ∞, you would be oblivious of what the x-coordinate of any finite point was.

Unfortunately, this idea of using ∞ as a coordinate isn't very good. Our usual arithmetic breaks down if we try to treat ∞ as a number. For instance, we'd have to say that $1 \cdot \infty = 2 \cdot \infty = \infty$. But if we could treat ∞ like a nonzero number (and it certainly isn't zero!), we should then be able to divide both sides of the equation by it, and deduce $1 = 2$.

So we've added one point at infinity in which all the lines parallel to L can intersect each other. If we start again with a line L' that is *not* parallel to L, then we've got to add another point at infinity, in which all lines parallel to L' can meet. We will see soon how to do this systematically, so that any two lines in the extended plane will meet in exactly one point. We call this extended plane the "projective plane."

There are various ways of getting a good handle on all these points at infinity that we are going to add to the plane. One way is to use

3-dimensional space and various geometric projections to systematize the concept of points at infinity. A nice exposition of this approach may be found in Courant and Robbins (1979, Ch. IV, §4). This use of projection is the reason the extended plane is called the "projective plane."

We will take a different approach in this book. Our attack will involve an algebraic method of defining the new points, one which is more suited to the algebraic explorations of the rest of the book.

> **IMPORTANT POINT**: Whichever approach we take, we only have to add *one* new point for each set of parallel lines. After that we will not have to add any more points. We will find that *any* line will intersect *any* curve, if we allow the intersection to be either a finite point or a point at infinity. In fact, *any* two *curves* will have intersection points in the projective plane. This beautiful fact is analogous to the fact that we only have to add i to the real numbers to obtain the algebraically closed complex numbers. We do not have to add more and more new "imaginary" numbers to solve more and more polynomial equations. On the other hand, when we extended the finite field $\mathbf{F}_p$ to an algebraically closed field $\mathbf{F}_p^{\text{ac}}$, we did have to keep adding new roots—there is no *single* polynomial whose roots, posited to exist alongside the original $\mathbf{F}_p$, will enable us to solve *all* polynomial equations with coefficients in $\mathbf{F}_p$.

2. Projective Coordinates on a Line

We are looking for an algebraic method that will add points at infinity to the plane. Let's begin by trying to add one point at infinity to a single line. Suppose we have a line L, with a coordinate t on it. We can think of L as a t-axis, and there needn't be any other axes around. So every point on L has a coordinate $t = a$, where a is some number. The brilliant idea is to think of the number a as a ratio. We can write any number as a fraction b/c. For example, $3 = 6/2$ and $0 = 0/5$. This may not seem so brilliant, but please be patient.

At the expense of a kind of numerical verbosity, we can designate the point on L with coordinate $t = a$ by the *two* "coordinates" (b, c) where

$a = b/c$. To distinguish the fact that (b, c) is going to designate a point on a line, not a point in a plane, we use a new notation: $(b : c)$. We call a symbol of the form $(b : c)$ "projective coordinates."

To repeat, the point designated by the projective coordinates $(b : c)$ is simply the point on the t-axis with coordinate $t = b/c$. What have we gained? We do not actually have to carry out the division. That leaves us free to contemplate a "point" that would be designated by the projective coordinates $(1 : 0)$. You are never allowed to divide by zero, so we don't attempt it. (If we did, we could say that this point is the point with $t = 1/0 = \infty$, but working with ∞ is exactly what we want to avoid.) We can convince ourselves that the "point" designated by the projective coordinates $(1 : 0)$ ought to be a new point, not on L, but added to it, and lying "at infinity."

Before we do this, we should mention two things. First, one potential drawback of this idea is that the same point will have lots of different projective coordinates. For example, the point with t-coordinate $t = 3$ can be designated with the projective coordinates $(6 : 2)$ or $(9 : 3)$ or $(3 : 1)$ or $(-15, -5)$ or $(1, 1/3)$. (Pay close attention to this last possibility, because we will use it in explaining why $(1 : 0)$ is the point at infinity.) This seems to be simply a price we have to pay for using projective coordinates. However, we will soon see it is the strength of the system. Second, we do not allow $(0 : 0)$ to be the projective coordinates of any point. You might think the point at infinity could be designated this way, but we will explain why not in a little while.

Let's take a trip to infinity along L, but let's do it in projective coordinates. For example, we could step from $(0 : 1)$ (where $t = 0$), to $(1 : 1)$ (where $t = 1$) to $(2 : 1)$ (where $t = 2$) and so on to $(k : 1)$ (where $t = k$) after our k-th step. As we let k get larger and larger, we step along toward "infinity," but there is no limiting value of k as a number. (Remember, we are disallowing any use of ∞ as a number.) What to do? We can record our trip in a different fashion. Keeping in mind the multiple choices we can employ for projective coordinates, these same points can be referred to as $(0 : 1)$, $(1 : 1)$, $(1 : 1/2)$, and on to $(1 : 1/k)$. Now as k gets larger and larger, the fractions $1/k$ tend nicely to 0. It makes sense to say that $(1 : 0)$ designates a point infinitely far out (to the right, if we think of L as a horizontal axis in the usual way) on L.

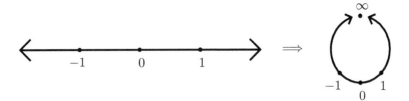

Figure 3.2. The real line and the projective real line

Perhaps we need a different point to stick on infinitely far to the left on L? Stepping to the left, we go from $t = 0$ to $t = -1$ to $t = -2$ to $t = -k$ after the k-th step. In projective coordinates we can write these points as $(0 : 1)$, $(1 : -1)$, $(1 : -1/2)$, and on to $(1 : -1/k)$. Once again, when k goes to infinity, $-1/k$ goes to zero. *We end up at the point with projective coordinates* $(1 : 0)$ *again*, whether we step to the right or the left!

This is why we only need to add one point to L. Pictorially, adding on the point at infinity must be performed both to the left and the right of the line, which means you must bend the line and glue the right and left ends of the line together. It's like inserting the right end into a tinker toy and the left end into the same tinker toy. What you get is a circle. (See figure 3.2.)

So far, we have only looked at points on L with real coordinates. If we allow complex coordinates, we could step off in other "directions." For example, we could look at points whose projective coordinates are $(0 : 1)$, $(1 : i)$, $(1 : i/2)$, and on to $(1 : i/k)$. Once again, when k goes to infinity, i/k goes to zero! The usual points of the complex "line" must be plotted on a plane, with a real axis and a complex axis. To create the projective complex line, we add one more point, with projective coordinates $(1 : 0)$ lying infinitely far out in *every* direction. Think about it: When you connect the infinitely distant "edges" up to the new point $(1 : 0)$, you get a sphere. (See figure 3.3.) This is called the *Riemann sphere*, and we will have a little more to say about it when we look at the complex points on elliptic curves in chapter 8.

To return to two things that we mentioned briefly already: First, the ability to change the projective coordinates by a scale factor—that is, to multiply both coordinates through by the same number and not change the point they designate (for example, $(1 : 1/k)$ and $(k : 1)$ designate the same point)—is exactly what allows us to see how to go off to "infinity" (in our example, to let k become larger and larger). In other words, all pairs

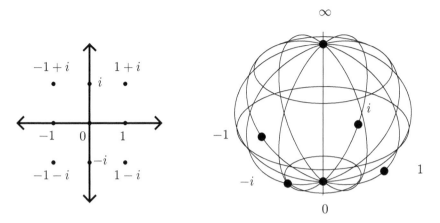

Figure 3.3. The complex plane and the Riemann sphere

of the form $(\lambda a : \lambda b)$ (as long as $\lambda \neq 0$) designate the *same* point in the projective line.

Second, why is $(0 : 0)$ no good for designating any point on the projective line? The first problem is that there is no point for it to match up with. Every point is already designated by a family of projective coordinate pairs. For instance, the point $t = 5$ can be designated by $(5a : a)$ for any nonzero number a. The point at infinity can be designated by $(a : 0)$ for any nonzero number a. All the points are already used up this way, and we don't need $(0 : 0)$ to designate anything. In addition, $(0 : 0)$ would be the odd man out: Because $(a \cdot 0 : a \cdot 0) = (0 : 0)$ for any a, the unfortunate point $(0 : 0)$ would be the only point with only one projective coordinate pair, and that would be strange.

Perhaps the most cogent reason that $(0 : 0)$ is no good for designating any point on the projective line, is that $(a : b)$ designates the point with $t = a/b$. This makes good sense as long as $b \neq 0$. And if we loosen up a bit, this even makes some sense if $b = 0$ as long as $a \neq 0$, because we can say (loosely) that $a/0 = \infty$. But the "ratio" $0/0$ makes no sense of any kind: It is totally ambiguous and could represent any and all numbers if it were to represent any. That's because if $0/0 = c$, then cross-multiplication gives the equation $0 = 0 \cdot c$. But this equation is true for all values of c, so c could be any number at all.

To summarize: The *projective line* consists of all points with projective coordinates $(a : b)$ where a, b can be any numbers, as long as not both are zero. Two pairs $(a : b)$ and $(a' : b')$ designate the same point if and only if $(a' : b') = (\lambda a : \lambda b)$ for some number $\lambda \neq 0$.

> **EXERCISE**: Group the following projective coordinate pairs so that those in each group designate the same point in the projective line, and different groups designate different points:

$$(0 : 1), (9 : 1), (0 : 2), (4 : 3), (3 : 4), (1 : \tfrac{1}{9}), (6 : 8), (3 : \tfrac{1}{3}), (12 : 9)$$

> **SOLUTION**: The pairs $(0 : 1)$ and $(0 : 2)$ represent the same point. The pairs $(9 : 1)$, $(1 : \tfrac{1}{9})$, and $(3 : \tfrac{1}{3})$ all represent the same point. The pairs $(4 : 3)$ and $(12 : 9)$ represent the same point. The pairs $(3 : 4)$ and $(6 : 8)$ represent the same point.

3. Projective Coordinates on a Plane

Now we can try to add points at infinity to the plane. Let's use the ordinary XY-coordinate system for the plane. Imitating what we did for the line, we can add another coordinate and convert to projective coordinates by looking at ratios. Calling the extra coordinate z, we write $(x : y : z)$ for the projective coordinates that designate the point P with regular coordinates $X = x/z$ and $Y = y/z$.

This makes good sense if $z \neq 0$, and P is a regular, old-fashioned point in the ("finite") plane. In particular, if $z = 1$, then $(a : b : 1)$ and (a, b) designate the same point in the plane, for any numbers a and b. But if $z = 0$, we are going to get new points "at infinity." These new points will be in the projective plane but not in the old, finite plane.

You might have thought of a different way of adding points at infinity to the plane. We could have brought in *two* new coordinates, one for X and one for Y. Suppose we call them z and w. We could then write $(x : z ; y : w)$, and hope that we have some kind of projective coordinate system for an extension of the finite plane. (Notice that we use a semicolon to separate the pairs of ratios $(x : z)$ and $(y : w)$. If we didn't do this, and if we considered all the ratios among all the four coordinates x, y, z, and w, we'd end up with some kind of projective 3-dimensional space, not a 2-dimensional plane.) In fact, the coordinate scheme $(x : z ; y : w)$ does define an extension of the finite plane, called $P^1 \times P^1$. Perhaps history could have been different, and $P^1 \times P^1$ would have been "the" projective plane. But that's not how history went, and we won't pursue this variant any more.

Mathematicians have agreed to consider the projective plane[2] to be the geometric system of points designated by the projective coordinates $(x : y : z)$. We should investigate

1. how many different points at infinity this commits us to;
2. where the points at infinity go; and
3. whether we will need more points at infinity than this for our improved theory of "geometric degree" to work for any algebraic curve.

Keep in mind that the point designated by the projective coordinates $(x : y : z)$ should depend only on the ratios between x, y, and z. If z is nonzero, then this point is "finite"; it is the regular old point (X, Y) with $X = x/z$ and $Y = y/z$. So the *new* points we are getting "at infinity" all occur when $z = 0$. As in the case we considered of adding the point at infinity onto a line, we will not allow the triple $(0 : 0 : 0)$ to be the projective coordinates of any point, finite or infinite.

So the new points that we are adding to the plane are those with projective coordinates that look like $(a : b : 0)$. But we do not have a 2-dimensional supply of new points, because only the *ratio* between a and b is important. If λ is nonzero, then $(\lambda a : \lambda b : 0)$ and $(a : b : 0)$ designate the same point in the projective plane. If you look back at what we did for the projective line, you will see that this set of new points has exactly the same form as what we did there, except for a third coordinate 0, which we are carrying along for the ride. In other words, to get from the usual plane to the projective plane, we are adding a whole "projective line" of points at infinity.

---

[2] We keep talking about *the* projective line and *the* projective plane. This terminology is correct only if we remember that we get different sets of points depending on the field we specify from which to draw our coordinate values. For example, there is the projective line with **R**-coordinates, called the "real" projective line. There is the projective line with **C**-coordinates, called the "complex" projective line. Those are the two instances we have been discussing. Another important instance for us will be the projective line with coordinates drawn from a finite field **F**. This is sometimes called a "finite" projective line, but we will avoid that, since we are using "finite" in this context to refer to the finite part of the line, that is, the garden-variety, old-fashioned, nonprojective line. The same comments apply as well to "the" projective plane.

To summarize, the projective plane consists of all points with projective coordinates $(a : b : c)$ where a, b, and c can be any numbers, as long as a, b, and c are not all 0. Two triples $(a : b : c)$ and $(a' : b' : c')$ designate the same point if and only if $(a' : b' : c') = (\lambda a : \lambda b : \lambda c)$ for some number $\lambda \neq 0$. This equivalence relation is called *homothety*.

EXERCISE: Group the following projective coordinate triples so that those in each group designate the same point in the projective plane, and different groups designate different points:

$$(0 : 0 : 1), \ (1 : 0 : 1), \ (2 : 2 : 4), \ (4 : 4 : 0), \ (1 : 1 : 2),$$

$$(0 : 0 : \pi), \ (\tfrac{2}{3} : 0 : \tfrac{2}{3}), \ (2 : -3 : 4),$$

$$(-2 : -2 : 0), \ (-3 : -3 : -6), \ (3\pi : 3\pi : 3\pi), \ (-4 : 6 : -8)$$

SOLUTION: The triples $(0 : 0 : 1)$ and $(0 : 0 : \pi)$ represent the same point. The triples $(1 : 0 : 1)$ and $(\tfrac{2}{3} : 0 : \tfrac{2}{3})$ represent the same point. The triples $(1 : 1 : 2)$, $(2 : 2 : 4)$, and $(-3 : -3 : -6)$ all represent the same point. The triples $(4 : 4 : 0)$ and $(-2 : -2 : 0)$ represent the same point. The triples $(2 : -3 : 4)$ and $(-4 : 6 : -8)$ represent the same point. The lonely triple $(3\pi : 3\pi : 3\pi)$ represents a point different from all of the others.

Let's see which of these points at infinity in the projective plane get added to any given line in the finite plane. Consider the parametrized line $(at + b, ct + e)$, where a and b are not both zero, and where t is the parameter. In projective coordinates, we could write this line as $(at + b : ct + e : 1)$. Now a point at infinity will have the projective coordinates $(x : y : 0)$. If we allow t to tend to infinity, when will $(at + b : ct + e : 1)$ tend toward $(x : y : 0)$?

Using our homothetic power to multiply through projective coordinates by nonzero numbers without changing the designated point, we can write the points along L as

$$\left(\frac{at + b}{ct + e} : 1 : \frac{1}{ct + e} \right).$$

(We ignore the value $t = -e/c$.) Assuming that $c \neq 0$, we see that as t tends to infinity, this expression tends to the triple $(a/c : 1 : 0)$. Boom! That's the point at infinity that sticks onto the "end" of L. You can check that this point depends only on L, not on the particular form of the parametrization you use to describe L. So in the projective plane, L is extended by one more point, at infinity. In this way L becomes a projective line on its own, which we could write $L \cup \{\infty\}$. (We start here a new convention of using the symbol ∞ loosely to stand for a point at infinity, to be determined from the context.)

To verify the claims above, let's start by using projective coordinates for L, and, to avoid confusion, use square brackets for the projective coordinates on L. Thus $[t : 1]$ designates the point on L with parameter value t. So $[t : 1]$ designates the same point as $(at + b : ct + e : 1)$. But we have learned to be more symmetrical, and to allow the last coordinate to vary also. Consider that $[t : s]$ designates projectively the same point $[t/s : 1]$, and this is the same point as $(a(t/s) + b : c(t/s) + e : 1)$. By the homothety rule, we can multiply these coordinates through by s, so this point is also designated by $(at + bs : ct + es : s)$. This is a nice symmetrical form. When $s = 1$, we get back where we started. But when we set $s = 0$, we get the point at infinity on L, namely the point designated by the coordinates $(at : ct : 0)$. You can check that this is the same as the answer we found before. If $s = 0$, we can be sure $t \neq 0$, so dividing through by t (which is the homothety by $1/t$), we get the coordinates $(a : c : 0)$, which designate this same point. If $c \neq 0$, we also get $(a/c : 1 : 0)$ designating this point, which is the form we found above. If $c = 0$, you'll have to divide through by a, which is left as an

EXERCISE: Carry out a similar analysis for the case when $c = 0$.

For the nonce, let's call a line through the origin $(0, 0)$ an "originary line." Then we see that in the projective plane, every finite line gets exactly one point at infinity added to it, so it becomes a projective line in the projective plane. Each point at infinity is added to exactly one originary line. And every line parallel to a given originary line gets that same point of infinity added on. To see this, think of the point at infinity $(f : g : 0)$. To what lines L does it attach to? Comparing with the results of our calculations, we see that it sticks onto lines parametrized by $(at + b, ct + e)$ where $a/c = f/g$. These are exactly the lines with slope c/a, so they are all parallel, and exactly one of them goes through the origin.

Our conclusion is particularly nice because one goal of our construction was to guarantee that all *parallel* lines get the *same* point of infinity added on to them, so that they all intersect in the *same* point in the projective plane.[3]

4. Algebraic Curves and Points at Infinity

The question now is: Suppose we have an algebraic curve C in the plane. Will it find points at infinity stuck on to it when we enlarge the plane to the projective plane? Will these points at infinity be enough to force the intersections we desire between the curve and the probing lines? The answers to these questions are "yes" and "yes."

We've already seen these answers in the case that C is a line. But now we need a systematic way to deal with *any* algebraic curve. Suppose C is given by the equation $f(x, y) = 0$, where $f(x, y)$ is a polynomial in two variables of degree $d \geq 1$. For example, let $f(x, y) = y - x^2$. Call the curve given by $y - x^2 = 0$, whose graph is a parabola, D.

Notice that naïvely speaking, it is not at all clear what points to stick onto D at infinity. There are no straight lines in D. The two arms of D are certainly going off to infinity, but are they tending off in some particular directions? Since the parabola gets steeper and steeper as you go off to the left or the right (see figure 3.4), you may think that both arms are tending to the vertical. Therefore, they should each run into that point at infinity that is at the end of all vertical lines. This intuition is correct: in the projective plane, the parabola closes up into a loop, with exactly one point at infinity; see figure 3.5.

---

[3] This makes it very hard to draw the real projective plane. In fact, it is not possible even to make a model of the real projective plane using rubber in 3-dimensional space, unless you allow the model to intersect itself. The real projective plane as embedded in 3-space is sometimes called a "cross-cap." If you paste two cross-caps together, you get a "Klein bottle," named after German mathematician Felix Klein (1849–1925). You can read about these things in a book about topology, or see pictures on the Internet. We will treat the projective plane algebraically, and not try to draw any pictures.

Figure 3.4. $y - x^2 = 0$

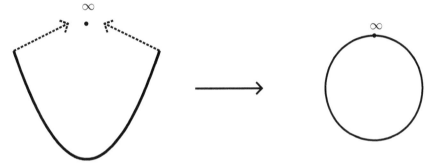

Figure 3.5. The projectivization of $y - x^2 = 0$

EXERCISE: Think about the hyperbola given by the equation $xy - 1 = 0$. How many points at infinity will it acquire in the projective plane? Will it close up into a loop? Into two loops? Into something else?

SOLUTION: The hyperbola $xy - 1 = 0$ has two asymptotes: the x-axis and the y-axis. It will acquire 2 points at infinity. It will share one of those points with the x-axis, namely $(1 : 0 : 0)$, and the other with the y-axis, namely $(0 : 1 : 0)$. The hyperbola will close up into a single loop.

Imagine traveling along the hyperbola in the positive x-direction, starting at $(1, 1)$. Eventually, your trip gets to a point at infinity, $(1 : 0 : 0)$, and then you will keep traveling back into the finite plane, now with negative x-coordinates. As you continue traveling, you will start to head down toward the negative part of the y-axis, and reach the other point at infinity on the hyperbola, $(0 : 1 : 0)$. Your trip will then continue by taking you into very large positive y-coordinates, and you will eventually end up back at $(1, 1)$.

5. Homogenization of Projective Curves

We need to develop a systematic algebraic method to answer which points at infinity (if any) to stick onto any given algebraic curve. The projective plane contains the finite plane, which consists of those points with projective coordinates $(x : y : 1)$, i.e., $(x : y : z)$ with $z = 1$. This is the key. If $z = 1$, we can multiply anything in sight by z without changing it. For example, our parabola D could also be expressed by the equations $yz - x^2 = 0$, or $yz^4 - x^2z^7 = 0$, or $y - x^2z = 0$, etc., as long as we fix $z = 1$. Of all of these possibilities, *only one makes sense under homothetic transformation.*

What does that italicized claim mean? Assume that $\lambda \neq 0$. Remember that $(x : y : z)$ and $(\lambda x : \lambda y : \lambda z)$ represent the same point in the projective plane. Now a function such as $G(x, y, z) = y - x^2z$ changes its values if we replace (x, y, z) by $(\lambda x, \lambda y, \lambda z)$. For example, $G(2, 1, 3) = 1 - 2^2 \cdot 3 = -11$ but $G(4, 2, 6) = 2 - 4^2 \cdot 6 = -94$. We can even find a λ so that G becomes 0: $G(2\sqrt{1/12}, \sqrt{1/12}, 3\sqrt{1/12}) = 0$. So the equation $G(x, y, z) = 0$ doesn't define anything if interpreted as an equation concerning points in the projective plane. If we tried to decide whether the point P designated by $(2 : 1 : 3)$ satisfied that equation, we'd get ambiguous answers, depending on whether we used $(2 : 1 : 3)$ or $(2\sqrt{1/12}, \sqrt{1/12}, 3\sqrt{1/12})$ to designate P.

But if we take $F(x, y, z) = yz - x^2$, something miraculous happens. It is true that the value of F at P still depends on which way we write the projective coordinates of P—so that the *value* $F(P)$ is meaningless, or not well-defined, and we won't talk about it. *Except* that it does make sense to say $F(P) = 0$ or $F(P) \neq 0$. Why? First look at our example: $F(2, 1, 3) = 1 \cdot 3 - 2^2 = -1$, and $F(4, 2, 6) = 2 \cdot 6 - 4^2 = -4$, and in general $F(\lambda \cdot 2, \lambda \cdot 1, \lambda \cdot 3) = 3\lambda^2 - 4\lambda^2 = \lambda^2(3 - 4) = -\lambda^2$, which is never zero, since $\lambda \neq 0$. So if P is the point designated by $(2 : 1 : 3)$, then it makes sense to say $F(P) \neq 0$, even though we cannot sensibly specify a value for $F(P)$.

On the other hand, $F(2, 4, 1) = 0$ and so does $F(2\lambda, 4\lambda, \lambda) = 0$ for any λ, because $F(2\lambda, 4\lambda, \lambda) = 4\lambda \cdot \lambda - (2\lambda)^2 = \lambda^2(4 - 4) = 0$. So if Q is the point designated by $(2 : 4 : 1)$, then it makes perfectly good sense to say $F(Q) = 0$, because this will hold regardless of which homothetic version of the projective coordinates for Q we use.

Now what is the difference between F and G that made things work for F? If you look at the last two paragraphs, you will see that we were able to factor out λ^2 from all terms when we evaluate $F(\lambda \cdot x, \lambda \cdot y, \lambda \cdot z)$. But when we evaluate $G(\lambda \cdot x, \lambda \cdot y, \lambda \cdot z)$, the two terms spit out different powers of λ, and then we can't factor anything out, so that these two terms can combine to give 0 or not give 0, depending on λ.

This happens because in the function $F(x, y, z) = yz - x^2$, both terms have the same degree, whereas in $G(x, y, z) = y - x^2 z$ the two terms have different degrees.

> **DEFINITION**: A polynomial is *homogeneous* if all its monomial terms have the same degree, which is then necessarily the degree of the polynomial.

For example, $x^n + y^n - z^n$ is homogeneous of degree n. But $x + y^2 + z^3$ is not homogeneous. Similarly, F was homogeneous of degree 2, but G was not homogeneous.

If $H(x, y, z)$ is homogeneous, then it makes sense to ask and answer the question: Given a point P in the projective plane, does $H(P)$ equal 0 or not equal 0? (It will *not* make sense to ask for the *value* of $H(P)$ if it is not equal to 0, unless H were a constant polynomial.)

Let's prove this: First, notice that if you have a term of degree d and you multiply all the coordinates by λ, then you multiply the term by λ^d. For example, the term $x^2 y^4 z$ has degree $2 + 4 + 1 = 7$, and $(\lambda x)^2 (\lambda y)^4 (\lambda z) = (\lambda^2 x^2)(\lambda^4 y^4)(\lambda z) = \lambda^7 x^2 y^4 z$. Now suppose P is designated by $(x : y : z)$ with fixed x, y, z. Write $H = \sum \text{terms}$, where each term has degree d. (The symbol $\sum$ stands for "sum of.") Then for any $\lambda \neq 0$,

$$
\begin{aligned}
H(\lambda x, \lambda y, \lambda z) &= \sum \text{terms } (\lambda x, \lambda y, \lambda z) \\
&= \sum \lambda^d \text{ terms } (x, y, z) \\
&= \lambda^d H(x, y, z).
\end{aligned}
$$

So the value of H is either always 0 or never 0, regardless of what nonzero value we give to λ.

CONCLUSION: If H is homogeneous of degree $d \geq 1$, we can sensibly speak of all points P in the projective plane for which $H(P) = 0$, and we call this the *projective curve* with equation $H = 0$. In this case, we would say the curve has *algebraic degree d*. We will see that this definition will coincide with the concept of geometric degree given by the probing line construction once we finish our work.

Now let's go back to a plane curve, such as the parabola given by $f(x, y) = y - x^2 = 0$. We can throw in z's without changing this equation when $z = 1$. But if we want the result to be usable to define a projective curve, we should throw in the z's so that the polynomial we end up with is homogeneous.

DEFINITION: If $f(x, y)$ is any nonzero polynomial, then the *homogenization* of $f(x, y)$ is the unique homogeneous polynomial $F(x, y, z)$ so that the degree of $F(x, y, z)$ is the same as the degree of $f(x, y)$ and $F(x, y, 1) = f(x, y)$. If C is the set of points with coordinates (x, y) for all solutions of $f(x, y) = 0$, we set $\overline{C}$ to be the set of points with projective coordinates $(x : y : z)$ for all solutions of $F(x, y, z) = 0$.

The process of finding $F(x, y, z)$ is called *homogenizing* the polynomial $f(x, y)$, and it's much simpler than it sounds from the complicated definition. You only need to change each monomial in $f(x, y)$ by multiplying it by the power of z needed to make it have degree equal to the degree of $f(x, y)$.

For example, the homogenization of $f(x, y) = y - x^2$ is $F(x, y, z) = yz - x^2$. The degree of F is the same as the degree of f, namely 2. Now try the following:

EXERCISE: What is the homogenization of

- $x^2y - y + 4$?
- $x^7 + y^2 + xy - y + 1$?
- $x^n + y^n - 1$
- $x^5 + x^2y^3 + y^5$?

SOLUTION: The homogenization of $x^2y - y + 4$ is $x^2y - yz^2 + 4z^3$. The homogenization of $x^7 + y^2 + xy - y + 1$ is $x^7 + y^2z^5 + xyz^5 - yz^6 + z^7$. The homogenization of $x^n + y^n - 1$ is $x^n + y^n - z^n$. The homogenization of $x^5 + x^2y^3 + y^5$ is itself, because $x^5 + x^2y^3 + y^5$ is already homogeneous.

If we have a polynomial $f(x, y)$ in two variables, of degree $d \geq 1$, which defines a curve C in the plane, then we can homogenize f to get $F(x, y, z)$, and F will define a curve $\overline{C}$ in the projective plane. Clearly, setting $z = 1$ will get back the curve C. In other words, C is the part of $\overline{C}$ that lies in the finite plane. The extra points we stick on when we go into the projective plane will be the set $\overline{C} - C$, that is, the set of all points in $\overline{C}$ that are not in C.

For example, let's go back to the parabola: $f(x, y) = y - x^2$. When we homogenize $y - x^2$, we get the polynomial of three variables $F(x, y, z) = yz - x^2$. Setting $F(x, y, z)$ equal to 0, we get the equation $yz - x^2 = 0$. The solutions when $z = 1$ are the points on the parabola in the finite plane: $(x : y : 1)$ where $y \cdot 1 = x^2$. The solutions when $z = 0$ are the points on the parabola that are points at infinity: $(x : y : 0)$ where $y \cdot 0 = x^2$. We see that at infinity, x must also equal 0, and then y can be anything nonzero (remember that $(0 : 0 : 0)$ is not the coordinate triple of any point). So the points at infinity on this parabola are the point(s) $(0 : y : 0)$ with $y \neq 0$. Remember that you can divide the projective coordinates through by any nonzero number and they represent the same point as before. So we see (by dividing through by the nonzero number y) that all the coordinate triples $(0 : y : 0)$ with $y \neq 0$ represent the same single point, which we can standardize by taking $y = 1$.

In summary, the *projectivized parabola* consists of the usual parabola in the finite plane, plus one more point at infinity. The finite points all have coordinates of the form $(t : t^2 : 1)$ and the point at infinity has the coordinates $(0 : 1 : 0)$. Notice that this last point is also the point at infinity on any vertical line. This bears out our intuition that the arms of the parabola are tending to the vertical. Since both arms attach onto this point at infinity, the complete projective parabola is a single loop. Look back again at figure 3.5 to see if the picture makes more sense now.

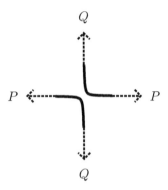

Figure 3.6. The projectivization of $xy - 1 = 0$

Another example: the hyperbola given by $f(x, y) = xy - 1 = 0$. Homogenizing $xy - 1$ gives $F(x, y, z) = xy - z^2$. The points on the hyperbola in the finite plane are given by setting $z = 1$, and have the projective coordinates $(x : y : 1)$ where $xy = 1$. The points at infinity on the hyperbola occur when $z = 0$. These will be the points $(x : y : 0)$ where $xy = 0$. This equation is solved when x or y or both are 0. But keep in mind that $(0 : 0 : 0)$ is not allowed to be the projective coordinate of a point. So x and y cannot both be 0 at the same time. We get points at infinity with projective coordinates $(x : 0 : 0)$, $x \neq 0$; and $(0 : y : 0)$, $y \neq 0$. In the first case, we can divide through x, and in the second case we can divide through by y, without changing the point being designated by the coordinates. We end up with 2 points at infinity on the hyperbola, namely those with projective coordinates $(1 : 0 : 0)$ (call it P) and $(0 : 1 : 0)$ (call it Q). The point P lies at infinity on all horizontal lines and Q on all vertical lines. P lies at infinity at the "end" of the horizontal asymptotes of the hyperbola, and Q at the "end" of the vertical asymptotes. See figure 3.6.

This enables us to review the solution to the exercise on page 55 with this notation. Follow these asymptotes around as if you were a bug crawling on the hyperbola. Suppose you started at the point A with $x = 1$, $y = 1$, and $z = 1$ and started crawling to the right. After a while you come to P at the "right end" of the horizontal asymptote. Where do you go from there? You do not want to retrace your steps, so you notice that P is also at the "left end" of the horizontal asymptote. So you can crawl in from the left and now you will move down along the hyperbola until you get to Q at the "bottom end" of the vertical asymptote. Continuing your trip, you move in from Q from the "top end" of the vertical asymptote, until you arrive back at the point A. You have completed a loop. Indeed the complete projective hyperbola, although it has two points at infinity, is a single loop, just like the projective parabola.[4]

In the next chapter, we will see how this all works when we probe a projective curve with a projective line. To make everything work right, we will have to count intersections with "multiplicities." Then in the following chapter, we will discuss Bézout's Theorem, which concerns counting all of the intersection points of two projective algebraic curves.

6. Coordinate Patches

Before we go on to study multiplicities, we want to mention a way to get a good look at what is going on near a point at infinity. To begin with an analogy, suppose you want to look at a map in an atlas of the world. Because the world is a sphere, and any page of the atlas is going to be flat and have edges, there is no way we can get an accurate mapping of the whole world onto one page. Each map will be centered at some point on the Earth. If you want to understand what's going on near you, you are best off finding a map that is centered at the point where you live. For example, if you live at the North Pole, you would like a map centered at the North Pole. But the South Pole probably won't be on the map at all, or if it is, the map will likely be distorted there. Your friends who live at the South Pole will use a map centered at the South Pole.

Now consider the projective plane. A part of the plane that we can study using just a pair of coordinates, like a page in an atlas, will be called a *coordinate patch*. When we contrast the coordinate pair in a coordinate patch to the projective coordinate triple, we will call the pair "affine coordinates" for that patch.

---

[4] There is a reason why both the projective hyperbola and the projective parabola consist of a single loop. There is a "projective change of coordinates" that transforms the parabola into the hyperbola, analogous to the ordinary change of coordinates that transforms a circle into an ellipse. In fact, the circle, the ellipse, the parabola, and the hyperbola are *all* equivalent under projective changes of coordinates.

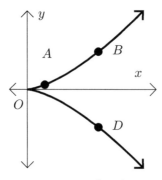

Figure 3.7. $C : x^3 - y^2 = 0$

For example, if you are interested in the usual, finite part of the projective plane, which is centered at the origin O with projective coordinates $(0 : 0 : 1)$, then you could just use the ordinary Cartesian coordinates (x, y). If you insist on using projective coordinates, then you may as well take $z = 1$ and use $(x : y : 1)$. Since the last coordinate is always 1, we can ignore it and just use the usual affine coordinates (x, y).

But if you are interested in what is happening near a point at infinity, then you might use a different "coordinate patch." For example, around the point P with projective coordinates $(1 : 0 : 0)$ (which is the point in which all horizontal lines converge), you could use the coordinates $(1 : y : z)$, or just the affine coordinates (y, z) for short. This coordinate patch again is an ordinary plane, but now it is the yz-plane, with origin P.

If you are studying a projective curve given by an equation $F(x, y, z) = 0$, where F is a homogeneous polynomial of degree d, then looking at the portion of the curve around O is accomplished by setting $z = 1$. This gives you the "dehomogenized" equation $F(x, y, 1) = 0$. But if you want to study this curve near P, you set $x = 1$, and get a different "dehomogenized" equation $F(1, y, z) = 0$. Note that the dehomogenized polynomial could have degree less than d. For example, the homogeneous polynomial $z^3 - xyz$ has degree 3, but upon setting $z = 1$, it dehomogenizes to $1 - xy$, which has degree 2.

For example, start with the curve C defined by $x^3 - y^2 = 0$. Then the projective curve $\overline{C}$, given by $x^3 - y^2 z = 0$ dehomogenizes around O (i.e., setting $z = 1$) to get $x^3 - y^2 = 0$, the equation that we started with. You can see what $\overline{C}$ looks like in the finite plane in figure 3.7.

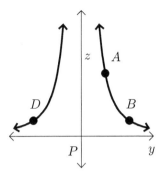

Figure 3.8. $C' : 1 - y^2z = 0$

If you want to look at $\overline{C}$ around P, dehomogenize $x^3 - y^2z = 0$ by setting $x = 1$, to get $1 - y^2z = 0$. Figure 3.8 shows what $\overline{C}$ looks like in a plane centered at P. We could call this curve C'.

We can figure out how C and C' fit together by considering the location of various points that are contained on both C and C'. For example, the point $A = (1 : 1 : 1)$ is easily seen to be on both curves, with the same affine coordinates on both, namely $(1, 1)$. The point $B = (4 : 8 : 1)$ is on C with coordinates $(4, 8)$, and on C' with coordinates $(2, 1/4)$. And the point $D = (4 : -8 : 1)$ is on C with coordinates $(4, -8)$ and on C' with coordinates $(-2, 1/4)$.

We can obtain a third coordinate patch centered at Q, the point with projective coordinates $(0 : 1 : 0)$, by setting $y = 1$ and using affine coordinates (x, z). This coordinate patch consists of all the points in the projective plane with y-coordinate nonzero. This gives the xz-plane, with origin Q. In this coordinate patch, the projective curve $\overline{C}$ becomes C'' defined by the equation $x^3 - z = 0$. The point A again has coordinates $(1, 1)$, the point B has coordinates $(\frac{1}{2}, \frac{1}{8})$, and the point D has coordinates $(-\frac{1}{2}, -\frac{1}{8})$. The result is in figure 3.9.

Notice that from the point of view of C, there is a point at infinity corresponding to the vertical "asymptote" of C as the arms of C flare upwards and downwards to the right. Algebraically, this is the point on $\overline{C}$ with $z = 0$, namely $(0 : 1 : 0)$, which we called Q. It appears clearly at the origin of C''.

If you are a bug wishing to take a voyage around the world on $\overline{C}$, starting at B, you can begin your journey by following your progress on C from B to A to O to D. When you arrive at D, you can switch coordinate patches

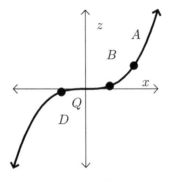

Figure 3.9. $C'' : x^3 - z = 0$

and follow your progress on C'' from D to Q and back to B. This is similar to driving across Europe and changing road maps as you move from one country to the next.

The three coordinate patches we have defined are called the "standard" coordinate patches. They are useful because they are obtained easily by setting one of the projective coordinates to 1, and between them they cover the entire projective plane.

EXERCISE: Let R be any point in the projective plane. Prove that R is contained in at least one standard coordinate patch.

SOLUTION: Let R have coordinates $(a : b : c)$. If a is nonzero, then R is in the standard coordinate patch defined by x nonzero. If $a = 0$ but b is nonzero, then R will be in the standard coordinate patch defined by y nonzero. If $a = b = 0$, then c *must* be nonzero, because $(0 : 0 : 0)$ cannot give the coordinates of any point in the projective plane. In this case, R is in the standard coordinate patch defined by z being nonzero.

So if you want to study what is happening at and near the point R (e.g., if you want to find out whether R is a singular point of a curve— a concept that we will define in the next chapter involving the lack of a unique tangent line to the curve at R), you may use one of these three standard coordinate patches to place R in that patch, and use the affine coordinates that go with that coordinate patch.

If you want to get more than one point in the same coordinate patch, then standard coordinate patches might not be enough. For example, the line at infinity is not wholly contained in any of our three standard coordinate patches. In fact, not even the two points with coordinates $(1 : 0 : 0)$ and $(0 : 1 : 5)$ are both contained in a single one of our three standard coordinate patches. (Which one could it be? The first point rules out y or z being nonzero, and the second point rules out x being nonzero.)

To fix this problem, we need more coordinate patches. We won't explicitly use these later in the book, but for completeness we'll tell you how to make them. If the field from which you are drawing coordinates is not a finite field, then any given finite set of points will be contained in one of these more general coordinate patches.

Here's what you do. Take three homogeneous linear polynomials of the variables, let's say $\xi(x, y, z) = ax + by + cz$, $\eta(x, y, z) = dx + ey + fz$, and $\zeta(x, y, z) = gx + hy + jz$. You have to do this so that the determinant

$$\det \begin{bmatrix} a & b & c \\ d & e & f \\ g & h & j \end{bmatrix} = aej + bfg + cdh - ceg - bdj - afh$$

is nonzero. You then define the coordinate patch to be all the points in the projective plane $(x : y : z)$ for which $\zeta(x, y, z) = gx + hy + jz \neq 0$. (Remember that you can multiply the coordinates $(x : y : z)$ through by a nonzero constant, and therefore the *value* of $\zeta(x, y, z)$ is not well-defined. But whether or not $\zeta(x, y, z)$ is nonzero is well-defined.) On this coordinate patch, the ratios $\xi(x, y, z)/\zeta(x, y, z)$ and $\eta(x, y, z)/\zeta(x, y, z)$ are well-defined, even if we replace (x, y, z) by $(\lambda x, \lambda y, \lambda z)$, and those ratios may be used as affine coordinates. The standard coordinate patch centered at $(0 : 0 : 1)$ is what results if we use $\xi(x, y, z) = x$, $\eta(x, y, z) = y$, and $\zeta(x, y, z) = z$.

As a final example in this chapter, let's use our machinery to find the point(s) at infinity on our model elliptic curve E given by $y^2 = x^3 - x$. The homogenized equation is

$$y^2 z = x^3 - xz^2.$$

We find the points at infinity on $\overline{E}$ by setting $z = 0$. This gives

$$0 = x^3$$

which implies that $x = 0$, and that in turn implies that $y \neq 0$. (Remember once again that $(0 : 0 : 0)$ is not the coordinates of any point in the projective plane.) There is just one point at infinity on $\overline{E}$, with projective coordinates $(0 : 1 : 0)$.

We will see later the importance of this point at infinity—it will be the identity element when we define a group law on $\overline{E}$.

. . . . .

Chapter 4

MULTIPLICITIES AND DEGREE

Road Map

The process of forcing our two definitions of degree to give
the same answer is still not at an end. We have already added
elements to our number systems and to the plane. Unfortunately, the line $y = 0$ will intersect the parabola $y = x^2$ only
at the point $(0,0)$, even if we use complex numbers and the
projective plane. To solve this problem, we must invent an idea
called *multiplicity*, which allows us to count certain points of
intersection more than once. In particular, we will count the
point $(0,0)$ in this example twice, forcing a solution to our
problem.

1. Curves as Varieties

If we have a homogeneous polynomial $F(x, y, z)$ in three variables, the
degree d of F is the degree of each and every one of the monomials in
$F(x, y, z)$. If we homogenize a polynomial $f(x, y)$ of two variables to get
$F(x, y, z)$, F and f will have the same degree. (Remember the connection:
$F(x, y, 1) = f(x, y)$.)

Each solution (x, y, z) to the equation $F(x, y, z) = 0$, when interpreted
as the projective coordinates $(x : y : z)$ of a point in the projective plane,
gives us the coordinates of a point P in that plane. The set of all those
points is the projective curve $\overline{C}$ defined by $F(x, y, z) = 0$. Those solutions
having $z = 1$ give us the coordinates of the part of the curve that lies in the
old-fashioned xy-plane. The set of those solutions is the curve C defined
by $f(x, y) = 0$. Those solutions having $z = 0$ give us the coordinates of

the part of the curve that lies "at infinity," consisting of some of the new points at infinity we added on to get the projective plane. The set of those solutions therefore is the complementary set $\overline{C} - C$.

Remember that every projective coordinate triple $(x : y : z)$ has the built-in homothetic property. That is, for any nonzero number λ, $(x : y : z)$ and $(\lambda x : \lambda y : \lambda z)$ designate the *same* point in the projective plane. Also remember that $(0 : 0 : 0)$ does not designate any point at all in the projective plane.

So far, we have not been specifying the field of numbers K we are using when we solve the equation $F(x, y, z) = 0$. (In our pictures, we have been assuming that $K = \mathbf{R}$.) The set of solutions naturally depends on the choice of the field K. If the field K is not understood from the context, then we will write $\overline{C}(K)$ for the set of solutions that we obtain when we draw the coordinate values from the field K.

Let's take $F(x, y, z) = x^2 + y^2 + z^2$ as an example, defining $\overline{C}$ with the equation $F(x, y, z) = 0$. We see that $\overline{C}(\mathbf{R})$ is the empty set, because the only solution in real numbers is $x = y = z = 0$, and $(0 : 0 : 0)$ does not designate any point in the projective plane. However, over the complex numbers,

$$\overline{C}(\mathbf{C}) = \left\{ (x : y : z) \mid z \text{ is a square root of } -x^2 - y^2 \right\}$$

contains infinitely many points. If we take a finite field, say $\mathbf{F}_2$, the field which contains only the two elements 0 and 1, solving $F(x, y, z) = 0$ by trial and error is easy. We can list the points, and we see that $\overline{C}(\mathbf{F}_2) = \{(1 : 1 : 0), (1 : 0 : 1), (0 : 1 : 1)\}$. (Remember yet again that $(0 : 0 : 0)$ does not designate any point in the projective plane.)

EXERCISE: List the points in $\overline{C}(\mathbf{F}_3)$. List each point only once. (For instance if you list $(1 : 1 : 1)$ and $(2 : 2 : 2)$, you have listed the same point twice.)

SOLUTION: The points are $(1 : 1 : 1)$, $(1 : 1 : 2)$, $(1 : 2 : 1)$, and $(1 : 2 : 2)$.

We did all this work because we are hoping to engineer the truth of the following assertion: Suppose $F(x, y, z)$ is a homogeneous polynomial of degree $d > 0$. If we intersect the solution set in the projective plane of $F = 0$ with a probing projective line $\overline{L}$, we will get *exactly d* intersection points—no matter which probe $\overline{L}$ we choose!

We have seen that to expect this, we had to expand the field of numbers from which we take the solutions to be algebraically closed, for example, $\mathbf{C}$ or $\mathbf{F}_p^{\mathrm{ac}}$. We have also seen that we have to expand the ordinary plane into the projective plane. There is one more thing we have to do: We have to count intersection points "with multiplicity." This is our next order of business.

2. Multiplicities

Start with a polynomial $p(x)$ in one variable of degree $d > 0$ and with coefficients in a field K, which we assume for the rest of this chapter is algebraically closed. Then $p(x)$ can be factored into linear factors, all with coefficients from K:

$$p(x) = c(x - a_1) \cdots (x - a_d),$$

and we can read off the solutions to $p(x) = 0$, namely $x = a_1, \ldots, a_d$.

How many numbers are on this list of solutions? It looks like d numbers, but some of them may be repeated. For example, suppose $p(x) = x^3 - 5x^2 + 7x - 3$ and the field $K = \mathbf{C}$. We factor p and obtain

$$p(x) = (x - 1)(x - 1)(x - 3).$$

So the solutions to $p(x) = 0$ are 1, 1, and 3. Or should we list them simply as 1 and 3? The first list contains more information and is preferable for many reasons. Listing the solutions with all repetitions is called listing the solutions "with multiplicity." In this example, we say that $x = 1$ is a solution to $p(x) = 0$ of multiplicity 2, and $x = 3$ is a solution to $p(x) = 0$ of multiplicity 1.

It is instructive to look at the graph of $y = p(x)$ in figure 4.1. The roots $x = 1$ and $x = 3$ can be found by locating where the graph intersects the x-axis. The difference in kind between the two roots is visible. At $x = 1$, the graph is tangent to the x-axis, while at $x = 3$, the graph slices the x-axis cleanly at a nonzero angle.

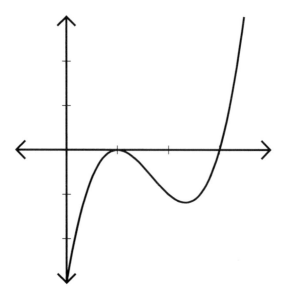

Figure 4.1. $y = (x-1)(x-1)(x-3)$

One obvious reason to list with multiplicities is that the total number of solutions (counted with multiplicity) will then be d, the degree of the polynomial whose roots we are finding. This observation is very important for us. Another reason is that in our example, the solution $x = 1$ in some sense is twice as important as the solution $x = 3$, and should be given its due in our reporting.

A third reason to count with multiplicities is to obtain "constancy of number." Consider another polynomial $\tilde{f}$ whose coefficients are just a little bit different from those of f. The practical amount of "a little bit" depends on the context. For example, consider the somewhat complicated polynomial $q(x) = x^3 - 5.2x^2 + 7.59x - 3.348$. Since degrees of q and p are the same, and the coefficients of q differ only a little from the coefficients of p, we consider q to be a "small perturbation" of p. When we factor q we get

$$q(x) = (x-0.9)(x-1.2)(x-3.1).$$

The solutions to $q(x) = 0$ are 0.9, 1.2, and 3.1. Now there are 3 distinct solutions, and we can describe this by saying that the "double root" 1 of $p(x)$ has split into the 2 roots 0.9 and 1.2 of $q(x)$ "under the perturbation."

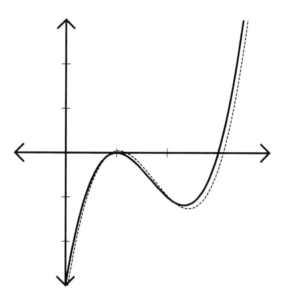

Figure 4.2. $y = (x - 1)(x - 1)(x - 3)$ and $y = (x - 0.9)(x - 1.2)(x - 3.1)$

We say that the number of roots of a polynomial is constant under perturbations, when we count with multiplicities. In this example, if we want to have constancy of the number of roots "near" 1, then we should count 1 twice as a root of p. Compare the graphs of $p(x)$ and $q(x)$ in figure 4.2.

What is a "small perturbation?" You can see that as we vary the roots of p more and more to get new polynomials, at some point we won't be able to say which roots are "near" 1 any more. We can continuously vary the roots until they become -20, -10, and 2. Which of these "came from" the double root of 1? It's a nonsensical question. We have to play it by ear. We'll only use vague words like "near" and "small" for descriptive purposes when it is clear what we are describing. Such words, without further specification, would not be allowed in official statements of theorems or rigorous proofs.

We summarize with a formal definition:

DEFINITION: Let $p(x)$ be a polynomial of degree $d \geq 1$ with coefficients in an algebraically closed field K. Let a be a number in K. The *multiplicity* of a as a root of $p(x)$ is m if and only if $p(x)$

factors as

$$p(x) = c(x - a)^m (x - b_1)(x - b_2) \cdots (x - b_{m-d})$$

for some $c \neq 0$, $b_1 \neq a$, $b_2 \neq a$, ..., $b_{m-d} \neq a$.

3. Intersection Multiplicities

Now let's think about a polynomial in two variables $f(x, y)$ and the curve C in the ordinary plane defined by $f(x, y) = 0$. Let L be a probing line in the plane. Then L and C may intersect in various points. Let's suppose they intersect in the point P with coordinates (ξ, η). We can define a "multiplicity" for this intersection by converting the study of the intersection into the study of a polynomial in a single variable and looking at the multiplicity of roots of this polynomial. How?

The number we will assign as the "multiplicity" of an intersection between C and L at the point P will be called the "local intersection multiplicity." Later we will add these local numbers up and talk about a "global intersection multiplicity." Local intersection multiplicities can also be defined for the intersections of two plane curves, but the exact definition in general is beyond the scope of this book.

What is the local intersection multiplicity of C and L at P? It's not hard to define. It will be defined to be a certain nonnegative integer, which we will call $I(C, L, P)$. To be completely general, if C and L don't actually meet at P, we'll say $I(C, L, P) = 0$.

From now on, let's assume that C and L do intersect at P. Parametrize the line L, say with parameter t, by $(x, y) = (at + b, ct + e)$, where at least one of a and c are nonzero. Then $P = (\xi, \eta)$ is a point on both the curve C and the line L. This means $f(\xi, \eta) = 0$ and there is some value of the parameter t so that $(at + b, ct + e) = (\xi, \eta)$. We can shift the parameter if we need to so that the line gets to P exactly at "time" $t = 0$. In other words, we may assume that $(b, e) = (\xi, \eta)$.

Now the intersections between L and C occur exactly for those values of t satisfying

$$f(at + b, ct + e) = 0.$$

Do you see why? Each point on L is of the form $(x, y) = (at + b, ct + e)$ for some t, and those coordinates (x, y) that make $f(x, y)$ vanish, are the points on C. Therefore, those points on L that are also on C are the points with coordinates $(at + b, ct + e)$ that make f vanish.

If you plug $x = at + b$ and $y = ct + e$ into the polynomial $f(x, y)$, you will get a new polynomial, call it $g(t) = f(at + b, ct + e)$, in the single variable t.

EXERCISE: Suppose that $f(x, y) = x^2 + y + 4$, and L is given by the parametric equations $x = 2t + 1$ and $y = t - 5$. Find the polynomial $g(t)$ in this case.

SOLUTION: When we substitute $x = 2t + 1$ and $y = t - 5$ into $f(x, y)$, we get

$$g(t) = (2t + 1)^2 + (t - 5) + 4 = 5t + 4t^2. \tag{4.1}$$

Note that when $t = 0$, the line L passes through the point $(1, -5)$, and we can check that $f(1, -5) = 0$. Also note that $g(t)$ has no constant term.

Going back to the general case, we'll get a polynomial $g(t)$ of some degree. This polynomial $g(t)$ must have a root at $t = 0$. Why? We have rigged the parametrization so that when $t = 0$, $(x, y) = (b, e) = (\xi, \eta)$ and we are assuming that $f(\xi, \eta) = 0$, since P is on C. So $g(0) = f(0 \cdot t + b, 0 \cdot t + e) = f(b, e) = f(\xi, \eta) = 0$.

Now that means that $g(t)$ has no constant term. It looks like

$$g(t) = r_0 t^k + r_1 t^{k+1} + r_2 t^{k+2} + \cdots + r_n t^{k+n}$$

for some $k > 0$ and $r_0 \neq 0$. In equation (4.1), for example, we have $k = 1$ and $r_0 = 5$.

We *define* $I(C, L, P) = k$ in this case. It's not hard to check that k does not depend on the choice of parametrization for L or the coordinate patch in which we work, so the definition is "well-defined." Notice the important fact that the solution $t = 0$ of $g(t) = 0$ has multiplicity k. This is because

$g(t)$ factors as

$$g(t) = t^k(r_0 + r_1 t^1 + r_2 t^2 + \cdots + r_n t^n) = r_n t^k \prod_{i=1}^{n}(t - q_i)$$

where all the q_i's are nonzero. (The product of the q_i's equals $(-1)^n r_0/r_n \neq 0$, as you can see easily by multiplying the factorization back out again.) Notice that the definition we just gave works no matter what field of numbers we use for the coordinates.

> **EXERCISE:** Find the local intersection multiplicity $I(C, L, P)$ for the field $K = \mathbf{C}$ in the following cases:
>
> (a) C is the curve $y^2 - x^3 - 3x = 0$, L is the line $y = 2x$, and $P = (1, 2)$.
> (b) C is the curve $y - x^3 = 0$, L is the x-axis, and $P = (0, 0)$.
> (c) C is the curve $y^2 - x^3 = 0$, $P = (0, 0)$, and L is any line through P.

SOLUTION: (a) We need to parametrize the line $y = 2x$ so that L passes through P when $t = 0$. One way to do this is to set $x = t + 1$ and $y = 2t + 2$. We substitute those two equations into $y^2 - x^3 - 3x = 0$, and we get $-t^3 + t^2 + 2t = 0$. Because the lowest exponent of t is 1, we have $I(C, L, P) = 1$.

(b) We can parametrize the x-axis with the pair of equations $x = t$ and $y = 0$. When we substitute those equations into $y - x^3 = 0$, we get $-t^3 = 0$. In this case, $I(C, L, P) = 3$.

(c) We parametrize the line L with the pair of equations $x = at$ and $y = bt$. Substitution of those two equations into $y^2 - x^3 = 0$ yields $b^2 t^2 - a^3 t^3 = 0$. Now, if $b \neq 0$, then we have $I(C, L, P) = 2$, while if $b = 0$ (in which case the line L is the x-axis), we have $I(C, L, P) = 3$. (Note that it is not possible for both a and b to be 0, for then we would not have a parametrization of a line.)

We may not always want to parametrize the line L so that L passes through the point P when $t = 0$. For example, we might want to study a case where L meets a curve in several different points $P_1, \ldots, P_m$ and we might want to study all of these points simultaneously. Therefore, we record the following useful fact:

PROPOSITION 4.2: Suppose C is defined by the equation $f(x, y) = 0$ and L by the parametrized line $(x, y) = (at + b, ct + e)$. Suppose the point P with coordinates (ξ, η) is on both L and C. That is, for some value t_0 of the parameter t, we have $(at_0 + b, ct_0 + e) = (\xi, \eta)$ and $f(\xi, \eta) = 0$. Then the equation $g(t) = f(at + b, ct + e) = 0$ has $t = t_0$ as a solution, and the multiplicity of the root $t = t_0$ of $g(t)$ is exactly equal to $I(C, L, P)$.

If you are good at algebra, you can prove this proposition.

We should drag out into the open something we have been hiding. The curve C defined by the equation $f(x, y) = 0$ is not merely the *set* of points P with coordinates (x, y) satisfying the equations. We have implicitly meant C to denote this set of points, together with the information that it is given by $f(x, y) = 0$ with a particular function $f(x, y)$. The points of C themselves can come with certain "multiplicities."

For a simple example of this, let $f(x, y) = x^2 + y^2 - 1$ and let the field we are dealing with be **R**. (In this paragraph, we relax the requirement that our field be algebraically closed.) Then the set of points on C, the curve defined by $x^2 + y^2 - 1 = 0$, is the circle of radius 1 centered at the origin. Next, consider a new polynomial $g(x, y) = (x^2 + y^2 - 1)^2$, defining a new curve C' by $(x^2 + y^2 - 1)^2 = 0$. The set of points on C' is exactly same circle, but somehow each point should be counted with multiplicity 2. You can see this by intersecting C and C' with the horizontal line L given by $x = t$, $y = 1$. Purely geometrically, in terms of points, L meets both C and C' at just the one point $(0, 1)$ when $t = 0$. But according to our definition of the local intersection multiplicity, $I(C, L, P) = 2$ while $I(C', L, P) = 4$. (Do the math: Plugging $x = t$, $y = 1$ into $f = 0$ gives $t^2 = 0$, and $t = 0$ is a root of multiplicity 2. Plugging $x = t$, $y = 1$ into $g = 0$ gives $t^4 = 0$, and $t = 0$ is a root of multiplicity 4.)

When we talk about an algebraic curve C, we are really referring to the equation $f(x, y) = 0$ that defines C. Although we will use geometry for explanatory or heuristic purposes, what we really mean will always be defined and computed algebraically.

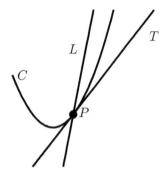

Figure 4.3. $I(C, L, P) = 1$

4. Calculus for Dummies

When is the local intersection number between a line and a curve greater than 1? We've seen we should count multiplicities, so sometimes what looks like just one point is mystically more than one. When?

There are two ways the local intersection number between a line and a curve can be greater than 1. To understand them, recall the concept of "tangency." A line T is tangent to a curve C at a point P if it touches C at P, and "near" P, if you "straighten out" C, you get the line T. (We'll give a rigorous algebraic definition of this concept later.) If L is a line that intersects C at P and $L \neq T$, then $I(C, L, P) = 1$. See figure 4.3 for an example. But if C does not have a unique tangent line at P (see figure 4.4) or if $L = T$ (see figure 4.5), then $I(C, L, P) > 1$. In both figures 4.4 and 4.5, you can see that if you move L a little bit, then L may intersect C in more than one point. That is not the case in figure 4.3. (The two curves in figure 4.4 are said to be "not smooth.")

In summary, the two ways that $I(C, L, P)$ can be greater than 1 are:

1. There can be something "wrong" with the curve, namely, it is not smooth.
2. There can be something "wrong" with the way the line meets the curve, namely, it could be tangent to the curve.

Perhaps "wrong" should be "wonderful" or "interesting." Mathematicians use a more neutral word, and call the curve "singular" in the first case, and the intersection "tangential" in the second case.

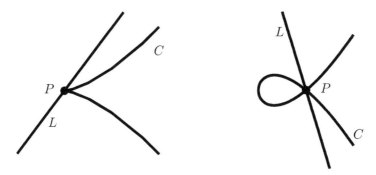

Figure 4.4. No unique tangent line

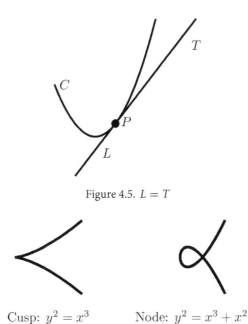

Figure 4.5. $L = T$

Cusp: $y^2 = x^3$ Node: $y^2 = x^3 + x^2$

Figure 4.6. Two types of singularities

It turns out both of these cases can be determined by looking at the tangent line (if any) to the curve at the point P. If the curve fails to have a well-defined tangent line at a point P, we say the curve is "singular" at P. (As you will see below, we only define a tangent line when the gradient is nonvanishing. So by our definition, a curve either has one tangent line at P, or is singular at P.) Two examples of singularities are in figure 4.6.

These are pretty pictures and show what is happening when the field we are using to solve these equations is the real numbers $\mathbf{R}$. But how can we draw the pictures when the field is the complex numbers $\mathbf{C}$? (We would need 4-dimensional paper.) And if the field is the algebraic closure of a finite field, such as $\mathbf{F}_7^{\mathrm{ac}}$, how can we see tangent *lines*?

If you think back to calculus, you may remember that you found the tangent line to a curve by differentiating. Differentiation of a function $f(x)$ is defined by a limit process, involving the difference quotient

$$\frac{f(x+h) - f(x)}{h}$$

and letting h tend to 0. This is great for $\mathbf{R}$, but can be tricky for $\mathbf{C}$. And for $\mathbf{F}_7$ the limiting process makes no sense at all.

The nice fact is that differentiating a *polynomial* function can be accomplished by means of an algebraic formula, without using any limiting. Remember the formula $\frac{d}{dx}(x^n) = (x^n)' = nx^{n-1}$? We just carry over this formula to any polynomial with coefficients in *any* field F:

DEFINITION: If $f(x) = a_0 + a_1 x + a_2 x^2 + \cdots + a_n x^n$, we define the *derivative* of $f(x)$ to be the polynomial $f'(x) = a_1 + 2a_2 x + \cdots + na_n x^{n-1}$. If $f(x)$ is the constant polynomial, we define its derivative to be 0.

That's all there is to it. Some surprising things can happen, especially for finite fields. For example, the derivative of the polynomial x^7 over the field $\mathbf{F}_7$ is $7x^6$. But over $\mathbf{F}_7$, 7 is equivalent to 0, so $7x^6 = 0x^6 = 0$. So the derivative of x^7 is zero over $\mathbf{F}_7$, even though x^7 is not the constant polynomial. (It's not even constant as a function on $\mathbf{F}_7$, since $0^7 = 0$ and $1^7 = 1$.)

We define partial derivatives in the same way. For example, if $f(x, y)$ is a polynomial in two variables, the *partial derivative* of f with respect to x, written f_x or $\frac{\partial f}{\partial x}$, is what you get by thinking of y as a constant and differentiating using the definition in the x-variable. The partial derivative of f with respect to y, similarly written as f_y or $\frac{\partial f}{\partial y}$, is what you get by thinking of x as a constant and differentiating using the definition but now in the y-variable. For example, $\frac{\partial}{\partial x}(x^2 y^2 + 6y) = 2xy^2$, because we treat y as if it were constant.

EXERCISE: Suppose that $f(x, y) = 3x^2 + 11xy - 27xy^3$. What are $f_x(x, y)$ and $f_y(x, y)$?

SOLUTION: We have $f_x(x, y) = 6x + 11y - 27y^3$ and $f_y(x, y) = 11x - 81xy^2$.

Now suppose $f(x, y) = 0$ defines an algebraic curve C, so that $f(x, y)$ is a polynomial of degree at least 1. Let $P = (a, b)$ be a point on this curve, so that $f(a, b) = 0$.

DEFINITION: We define the *tangent line* to C at P to be the line given by the equation

$$f_x(a, b)x + f_y(a, b)y = e,$$

where $e = f_x(a, b)a + f_y(a, b)b$.

This choice of e ensures that the line goes through P. This is the line through P whose defining linear function has the same gradient at P as f. The *gradient* of f at P is the vector $(f_x(a, b), f_y(a, b))$ and the gradient of the line given by the equation $Ax + By = e$ is (A, B). We learn in calculus that the gradient is perpendicular to the tangent, and therefore the line whose gradient is the same as the gradient of C will be tangent to C.

The definition is pretty complicated, but after we do an example or two, applying it will become routine. For example, let's find the tangent line to the curve $f(x, y) = y^2 - x^3 - 3x = 0$ at the point $(0, 0)$. First the partial derivatives: $f_x = -3x^2 - 3$ and $f_y = 2y$. At the point $(0, 0)$, we have $f_x = -3 \cdot 0^2 - 3 = -3$ and $f_y = 2 \cdot 0 = 0$. So the tangent line is given by the equation $-3x + 0y = e$ where $e = (-3 \cdot 0 + 4 \cdot 0) = 0$. That is, the tangent line is given by the equation $-3x = 0$. You can see the result in figure 4.7.

We've drawn the picture for **R**, but the algebra we performed is valid for any field. If the field is $\mathbf{F}_5$, then $-3 = 2$ in that field and we get the equation $2x = 0$. If the field is $\mathbf{F}_3$, then $-3 = 0$ and we get the equation $0x + 0y = 0$ and WAIT! that is not the equation of a line—what is going on???

What just happened could happen even over the real numbers. For example, look at the curve given by $f(x, y) = y^2 - x^3 = 0$, pictured in

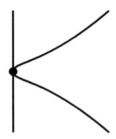

Figure 4.7. The tangent line to an elliptic curve

figure 4.6. Compute: $f_x = -3x^2$ and $f_y = 2y$. The "tangent line" at the origin, where $x = y = 0$, would be given by $0x + 0y = 0$. This is not the equation of a line.

So there may be points on a curve on which the tangent line is not defined, because $f_x = f_y = 0$ at those points. We call such points *singular points* of the curve. If a curve has singular points, we call it a *singular curve*. If it doesn't have any singular points, even over an algebraic closure of the field we are working with, we call it *nonsingular* or *smooth*.

It's quite likely for a curve to be nonsingular. Consider the curve given by $f(x, y) = 0$. A singular point P would have coordinates (a, b) where $f(a, b) = 0$, $f_x(a, b) = 0$, and $f_y(a, b) = 0$. These are *three* equations in only the *two* "unknowns" a and b, so for there to be any simultaneous solutions, something a bit special has to happen. Elliptic curves are nonsingular cubic curves.[1]

If P is a singular point on the curve C and L is any line meeting C at P, then the local intersection number $I(C, L, P)$ is always greater than 1. If you like algebra, you could try to prove this.

The other way that $I(C, L, P)$ can be greater than 1 is if L is tangent to C at P. By definition, this means that P is a nonsingular point for C and L is its tangent line. Again, if you like algebra, you could try to prove this.

---

[1] It can happen that a curve, in our sense of whatever is defined by the equation $f(x, y) = 0$, can have *every* point on it singular! Let $f(x, y) = x^2 - 2xy + y^2$. Then $f_x = 2x - 2y$, $f_y = -2x + 2y$, and $f(x, y) = (x - y)^2$. The points of the curve defined by the equation $f(x, y) = 0$ are those points where $x = y$, and at all of those points $f_x = f_y = 0$. More generally, consider the case where $f(x, y) = g(x, y)^2$, for some polynomial g of degree at least 1. The usual laws of calculus hold for our definition of derivatives of polynomials, including the sum rule, the product rule, and the chain rule. So $f_x = 2gg_x$ and $f_y = 2gg_y$. If P is any point on the curve, then $f(P) = 0$. Therefore, $g(P) = 0$, and therefore $f_x(P) = 0$ and $f_y(P) = 0$. What's happening here is that the whole curve has multiplicity 2 (assuming g is an irreducible polynomial).

Parametrize L so that the point P corresponds to $t = 0$, and substitute that parametrization for L into the equation for C. Because L is tangent to C at P, $t = 0$ will be a root of multiplicity at least 2.

In the next chapter, we will see that we have accomplished our goal in Part I: If we count the points of intersection of any line with an algebraic curve of degree d in a projective plane over an algebraically closed field, counting the points with multiplicity, then there will be exactly d of them.

Chapter 5

· · • · ·

BÉZOUT'S THEOREM

Road Map

To show how powerful our new definitions are, we describe Bézout's Theorem, which counts all the intersections of two algebraic curves keeping in mind their multiplicities. We sketch a proof of the theorem in the case when one of the two curves is a line. In this case, the theorem justifies our claim in the last paragraph of the previous chapter.

The details of the proof do not recur again in our journey, but their beauty justifies our small excursion.

1. A Sketch of the Proof

Sometimes, we can't get everything that we want, and sometimes we want something general rather than something specific. For example, let's suppose that we want to solve the system of equations

$$\begin{cases} f(x, y) = 0 \\ g(x, y) = 0 \end{cases}$$

where $f(x, y)$ and $g(x, y)$ are polynomials. It could be very hard to find a list of solutions, but perhaps we would settle for knowing the number of solutions. Maybe there is a general theorem telling us how many solutions there are in terms of a simple rule that we can apply to f and g.

The preceding chapters have hinted that this particular problem won't have a nice answer unless we look for all of the answers in an algebraically closed field, we include solutions "lying at infinity," and we count solutions with multiplicity. If we do all of this, then there is a nice answer to the problem of counting the number of solutions, called Bézout's Theorem. (Étienne Bézout was a French mathematician who lived from 1730 to 1783.) We will discuss the answer in terms of intersection numbers. There is a way to define a local intersection number $I(\overline{C}, \overline{D}, P)$ where $\overline{C}$ and $\overline{D}$ are *any* two projective algebraic curves in a projective plane, and P is a point where they intersect. The details of the definition go beyond the scope of our book. But once the details can be handled, we can formulate a new definition:

DEFINITION: Suppose that $\overline{C}$ and $\overline{D}$ are 2 projective curves, such that $\overline{C} \cap \overline{D}$ contains only a finite number of points $P_1, P_2, \ldots, P_n$. The *global intersection multiplicity* of $\overline{C}$ and $\overline{D}$ is
$\mathscr{I}(\overline{C}, \overline{D}) = \sum_k I(\overline{C}, \overline{D}, P_k)$.

That is to say, the global intersection multiplicity is just the sum of all of the local intersection multiplicities over all of the points of intersection, including of course the points at infinity.

For this chapter, we assume that K is an algebraically closed field, and that all of the polynomials we consider have coefficients in K.

BÉZOUT'S THEOREM: Assume that the intersection of $\overline{C}$ and $\overline{D}$ consists of finitely many points. Then $\mathscr{I}(\overline{C}, \overline{D})$ is exactly the product of the degrees of the homogeneous polynomials defining $\overline{C}$ and $\overline{D}$.

Our goal in this section is to sketch a proof of Bézout's Theorem when $\overline{D}$ is a projective line $\overline{L}$. Bézout's Theorem in this case just says that the sum of all the local intersection numbers $I(\overline{C}, \overline{L}, P)$ over all points P of intersection equals the degree of the polynomial defining $\overline{C}$, because the degree of the polynomial defining $\overline{L}$ is 1. This proof will then show the equality of our two concepts of degree of a curve:

1. The degree of the polynomial defining the curve (algebraic degree).
2. The number of points of intersection of any probing line with the curve (geometric degree).

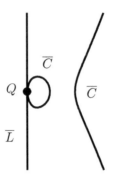

Figure 5.1. $y^2 = x^3 - x$ and $x = -1$

Our whole labor up to now has been to explain how to count the number of points of intersection. We have to be careful to look at all points with coordinates in some fixed *algebraically closed field*, we have to be careful to look at all points, *including those at infinity*, in the projective plane, and we have to be careful to count local intersections *with multiplicity*.

You can skip the following proof-sketch without impairing your understanding of the rest of this book. If you do read it, remember that proofs usually become clearer if they are read and re-read in small chunks. Even if you don't read it right now, you can have a look at this example, based on our model elliptic curve:

Let $\overline{C}$ be given by $y^2 z = x^3 - xz^2$, and let $\overline{L}$ be given by $x + z = 0$. In figure 5.1, we show the pieces of $\overline{C}$ and $\overline{L}$ gotten by setting $z = 1$. Remember that the point at infinity, $P = (0 : 1 : 0)$, is also part of both $\overline{C}$ and $\overline{L}$.

The degree of $\overline{C}$ is 3, and the degree of $\overline{L}$ is 1, so by Bézout's Theorem, $\overline{C}$ and $\overline{L}$ should intersect with global intersection multiplicity equal to $3 \cdot 1 = 3$. Let's see how the computation proceeds:

We can see that $\overline{C}$ and $\overline{L}$ intersect at $Q = (-1 : 0 : 1)$, and they also intersect at the point $P = (0 : 1 : 0)$. We claim that these are the only two intersection points, even if we use complex numbers. To check, solve

$$\begin{cases} y^2 z = x^3 - xz^2 \\ x + z = 0 \end{cases}$$

by setting $x = -z$ and substituting into the first equation. We get $y^2 z = (-z)^3 - (-z)z^2 = -z^3 + z^3 = 0$, and therefore either $y = 0$ or $z = 0$.

If $y = 0$, then we have $x = -z$, and we get the solution $(-1 : 0 : 1)$. Because of homothety, this is the only projective solution with $y = 0$.

If $z = 0$, the equation $x = -z$ forces $x = 0$, and we get the solution $(0 : 1 : 0)$. Again, because of homothety this is the only projective solution. So the global intersection number is

$$\mathscr{I}(\overline{C}, \overline{L}) = I(\overline{C}, \overline{L}, P) + I(\overline{C}, \overline{L}, Q).$$

To compute $I(\overline{C}, \overline{L}, P)$, we need a coordinate patch centered at P. Take $y = 1$, and that gives the affine curves $z - x^3 + xz^2 = 0$ and $z = -x$. Those two curves intersect at P with a local intersection number of 1. Why? Parametrize the line with

$$x = -t$$
$$z = \ \ \ t$$

so that the line L passes through P when $t = 0$. Substitute those two equations into the cubic equation, and we have the equation $t = 0$, and this has a root of multiplicity 1. Therefore, $I(\overline{C}, \overline{L}, P) = 1$.

To compute $I(\overline{C}, \overline{L}, Q)$, we can use the coordinate patch $z = 1$. Parametrize the line L with

$$x = -1$$
$$y = \ \ \ t$$

which passes through Q when $t = 0$. Substitute those two equations into the cubic $y^2 - x^3 + x = 0$ and we get the equation $t^2 = 0$, which has 0 as a root of multiplicity 2. Therefore, $I(\overline{C}, \overline{L}, Q) = 2$.

So $\mathscr{I}(\overline{C}, \overline{L}) = I(\overline{C}, \overline{L}, P) + I(\overline{C}, \overline{L}, Q) = 1 + 2 = 3 = 3 \cdot 1 = \text{degree}(\overline{C}) \cdot \text{degree}(\overline{L})$, as predicted by Bézout's Theorem.

SKETCH OF A PROOF OF BÉZOUT'S THEOREM: Let's assume that the curve $\overline{C}$ is defined by $F(x, y, z) = 0$, where $F(x, y, z)$ is a homogeneous polynomial of degree $D \geq 1$ with coefficients in K. Let $\overline{L}$ be the projective line given by the parametrization

$P(t : s) = (at + bs : ct + ds : et + fs)$, with a, b, c, d, e, and f in K. This can be thought of as a function from the projective line with coordinates $(t : s)$ to the projective plane. As such, the coefficients a, b, c, d, e, and f are assumed to have the property that $(at + bs : ct + ds : et + fs)$ can never equal $(0 : 0 : 0)$ unless both t and s equal 0.

The intersection points of $\overline{C}$ and $\overline{L}$ are those points $P(t : s)$ such that $F(at + bs, ct + ds, et + fs) = 0$. Now this last polynomial $G(t, s) = F(at + bs, ct + ds, et + fs)$ in t and s has degree D also, unless it is identically 0. Why? Just look at any term of F: $\lambda x^u y^v z^w$ where λ is in K and $u + v + w = d$. Plugging in $x = at + bs$, $y = ct + ds$, and $z = et + fs$ we get $\lambda(at + bs)^u(ct + ds)^v(et + fs)^w$. We get many terms in t and s, but, by the binomial theorem, each one has degree D. So every term of $G(t, s)$ has degree D and thus so will $G(t, s)$ itself, unless all these terms cancel out and G is the zero polynomial.

If G is identically 0, that means that the whole line $\overline{L}$ is contained in the curve $\overline{C}$. The hypothesis of Bézout's Theorem says that $\overline{L}$ and $\overline{C}$ do not meet in infinitely many points. So we may assume that the degree of G is exactly D.

Notice the very important fact that only in projective geometry can we know that G has degree D. If we try to do this same reasoning with a nonhomogeneous polynomial $f(x, y)$ to start with, all the terms of degree D could cancel out without *all* the terms canceling. So $f(at + b, ct + e)$ could be nonzero, but of degree less than the degree of f. This is exactly why we had to extend our plane to be the whole projective plane in order to make the probing line intersect at enough points.

What are the points of intersection P of $\overline{C}$ and $\overline{L}$? They correspond to values of $(t : s)$ such that $G(t, s) = 0$. As usual in projective coordinates, parameter values $(t : s)$ and $(\lambda t : \lambda s)$ refer to the same point P, for any $\lambda \neq 0$.

Here's the sketchy bit. It has to be checked that the local intersection multiplicity $I(\overline{C}, \overline{L}, P)$ at the point P is given by the following rule: First, one proves that $I(\overline{C}, \overline{L}, P)$ doesn't depend on how we parametrize L. Second, one proves that we can always choose our parametrization of $\overline{L}$ so that

- $e = 0$, and
- if $\overline{L}$ is not the line at infinity, then the point at infinity on $\overline{L}$ is given by $s = 0$.

This choice of parametrization greatly simplifies the remainder of the proof. First, suppose that $\overline{L}$ is not the line at infinity. At a point of intersection in the ordinary plane, for which $s \neq 0$, you just set $s = 1$ and get the polynomial $h(t) = G(t, 1)$. Factor $h(t)$ into linear factors and the local intersection multiplicity $I(\overline{C}, \overline{L}, P)$ for the point corresponding to $t = t_0$ is exactly equal to the exponent of $(t - t_0)$ in this factorization.

For a point of intersection at infinity, you set $t = 1$ and get the polynomial $j(s) = G(1, s)$. (Notice that if a given line intersects the line at infinity in two or more points, then it *is* the line at infinity. This is because if two lines meet in at least two different points, then they are the same line. So there can be at most one point of intersection at infinity, namely when $s = 0$.) You then factor $j(s)$ into linear factors and the local intersection multiplicity $I(\overline{C}, \overline{L}, P)$ for the point corresponding to $s = 0$ (which is the point at infinity on the line) is exactly equal to the exponent of $(s - 0) = s$ in this factorization.

Now look at $G(t, s) = a_0 s^D + a_1 s^{D-1} t + a_2 s^{D-2} t^2 + \cdots + a_D t^D$ for some numbers $a_0, \ldots, a_D$ in K. Setting $s = 1$ gives $h(t) = a_0 + a_1 t + a_2 t^2 + \cdots + a_D t^D$. The degree of this polynomial is u, where u is the largest integer such that $a_u \neq 0$. Therefore if we sum the intersection multiplicities over all the points of intersection where $s = 1$ and t runs through the various t_0's, we get a total of u. This is because a polynomial of degree u has exactly u roots in an algebraically closed field, counted with multiplicity.

On the other hand, setting $t = 1$ gives $j(s) = a_0 s^D + a_1 s^{D-1} + a_2 s^{D-2} + \cdots + a_D = a_0 s^D + \cdots + a_u s^{D-u}$. The largest power of s that can be factored out of $j(s)$ is s^{D-u}. So the intersection multiplicity at the point given by $t = 1$ and $s = 0$ is $D - u$. Bézout's Theorem (in the case of a probing line which is not the line at infinity) now follows from the simple equation $u + (D - u) = D$.

Finally, let's do the case where $\overline{L}$ is the line at infinity. We could parametrize it by $(t : s : 0)$ and get $G(t, s) = F(t, s, 0)$. Then we could work out the rules for the local intersection multiplicities and check Bézout's Theorem works out, similarly to what we just did.

However, an easier method is to switch to a different coordinate patch, say the one where $y = 1$. In the xz-coordinates, $\overline{L}$ is no longer the line at infinity. Then our previous analysis holds good and finishes the proof. $\qquad\qquad\qquad\qquad\qquad\qquad\qquad\qquad\qquad\qquad\qquad$ $\square$

2. An Illuminating Example

Let's see how this goes in another example over the complex numbers, where we try several different probing lines. Let's look at the intersection of the parabola $\overline{C}$ given by $F(x, y, z) = yz - x^2 = 0$ with probing lines, which together will exemplify the 3 extensions (algebraic closure, projective plane, and multiplicity) that we have introduced. In each case, the sum of all the local intersection numbers should be 2, the degree of F.

First, let's take the probing line $\overline{L}_1$ given by $(t : -s : s)$. Substituting $(x, y, z) = (t, -s, s)$ in $yz - x^2 = 0$ we get $-s^2 - t^2 = 0$. The intersection points in the ordinary plane can be found when $s = 1$. The affine part of $\overline{C}$ comes from setting $z = 1$ and is the parabola C with equation $y = x^2$. The affine part of $\overline{L}_1$ is the horizontal line with equation $y = -1$. The intersection occurs when $-1 - t^2 = 0$, and we get the 2 solutions $t = \pm i$. (Note that we must use the algebraically closed field $\mathbf{C}$ to get these points.) These two intersection points are thus the points with projective coordinates $(\pm i : -1 : 1)$, and each has multiplicity 1, since $t^2 + 1$ factors as $(t - i)(t + i)$, where obviously each factor has multiplicity 1. The intersection points at infinity in the projective plane can be found when $s = 0$. This gives $t^2 = 0$, and we get the solution $s = t = 0$, but this would correspond to a point with projective coordinates $(0 : 0 : 0)$ and there is no such point. So there are no intersection points at infinity, and 2 in the ordinary plane, each of multiplicity 1, and $1 + 1 = 2$.

Second, let's take the probing line $\overline{L}_2$ given by $(0 : t : s)$. Dehomogenize, again setting $z = 1$ and $s = 1$. In the ordinary plane, we are now dealing with the same parabola C whose equation is $y = x^2$. Now the vertical line L_2 has equation $x = 0$. There is one intersection point within the complex numbers, $x = 0$ and $y = 0$. The local intersection number is 1, since substituting the line $x = 0$, $y = t$ into the equation $y = x^2$ gives $t = 0$, and this polynomial has a root of multiplicity 1.

The intersection points at infinity in the projective plane can be found when $s = 0$. Then t can be anything nonzero. (The value of t must be nonzero, because $(0 : 0 : 0)$ is not the projective coordinates of any point.) We get the solution $s = 0$ and $t = k$ (where k is any nonzero number). These solutions correspond to the point with projective coordinates $(0 : k : 0)$. Although k is arbitrary, these triples all designate the same point in the projective plane. We may as well take $k = 1$, so this intersection point at infinity is the one with coordinates $(0 : 1 : 0)$.

We find the local intersection multiplicity at a point at infinity as we did in the elliptic curve example. We have to use a coordinate patch in which the point of intersection at infinity becomes an ordinary point. We can do this by setting $y = 1$ and looking in the xz-plane. The equation of our parabola now becomes $z - x^2 = 0$. The point we are worried about, with projective coordinates $(0 : 1 : 0)$ corresponds to $x = 0$, $z = 0$. The line L_2, which projectively is parametrized by $(0 : t : s)$, in xz-coordinates is given by $(0, s)$ (since we have set $y = t = 1$). Plugging $x = 0$ and $z = s$ into $z - x^2 = 0$, we get the equation $s = 0$. Again, this equation only has a root of multiplicity 1, and so the local multiplicity of this intersection is 1.

In conclusion, there is one intersection point at infinity, and one in the ordinary plane, each of multiplicity 1, and $1 + 1 = 2$. Check. Note that without the points at infinity, we wouldn't have been able to fill up our quota of points, even using complex numbers.

Third, let's take the probing line $\overline{L}_3$ given by $(t : 0 : s)$. Dehomogenize, again setting $z = 1$ and $s = 1$. In the ordinary plane, we are now dealing with the same parabola C whose equation is $y = x^2$, and now the horizontal line L_3 whose equation is $y = 0$. There is one intersection point with real coordinates, the solution to $x^2 = 0$. This has the unique solution $x = 0$, but because the factorization is $x^2 = (x - 0)^2$ with exponent 2, the multiplicity of the root is 2 and therefore the local intersection number is 2. In this case, $\overline{L}_3$ is tangent to the parabola, and we have to count the intersections with multiplicity to get the right answer.

We leave it to you to check that $\overline{L}_3$ and the parabola do not have any points at infinity in common. So there is just one intersection point in the ordinary plane, of multiplicity 2, and none at infinity, and $2 = 2$. Check.

EXERCISE: Let $\bar{L}_4$ be the probing line corresponding to the horizontal line given by $y = 1$. Write a projective parametrization of $\bar{L}_4$, and find all of its local intersection multiplicities with the parabola. Check that the local intersection multiplicities sum to 2.

SOLUTION: The homogenization of L_4 is the equation $y = z$, and so the homogeneous coordinates on $\bar{L}_4$ are given by $(t : s : s)$. Setting $z = 1$ and $s = 1$, we get $t^2 = 1$, with the 2 solutions $t = \pm 1$ and local intersection multiplicities 1 at each of those 2 points.

Checking for points at infinity, we must set $s = 0$, and $y = z = 0$, resulting in $t = 0$. But the point $(0 : 0 : 0)$ is not permitted in projective space, so there is no intersection at infinity.

EXERCISE: Do the example of the same parabola and 3 lines, but now take all coordinates from the field with 2 elements $\mathbf{F}_2$. You should not need the algebraic closure of $\mathbf{F}_2$, because in this example all of the points of intersection have coordinates in $\mathbf{F}_2$.

SOLUTION-SKETCH: The parabola is given by $yz - x^2 = 0$. $\bar{L}_1$ is given by $(t : -s : s)$. $\bar{L}_2$ is given by $(0 : t : s)$. $\bar{L}_3$ is given by $(t : 0 : s)$.

In the case of $\bar{L}_1$, we have to solve $-s^2 - t^2 = 0$ and since $1 + 1 = 0$ in $\mathbf{F}_2$, we get the one solution $s = t = 1$. To find the local multiplicity at $(1 : 1 : 1)$ (don't forget that $1 = -1$ in $\mathbf{F}_2$), set $z = 1$ and $s = 1$, and get the equation $1 + t^2 = 0$. This factors as $(t - 1)^2 = 0$, and so has multiplicity 2. The only intersection point is when $s = t = 1$, and the coordinates are $(1 : 1 : 1)$. The local multiplicity is 2. There is no intersection at infinity.

In fact, over the field $\mathbf{F}_2$, $\bar{L}_1$ is tangent to the parabola at the point of intersection. To see this, set $f(x, y) = y - x^2$ and compute $f_x = 0$ and $f_y = 1$, so the tangent line to any point on the parabola is given by $y = k$ for some constant k. For the tangent line to go through the point $(x, y) = (1, 1)$, we must have $k = 1$. So the tangent line to the parabola at $(1, 1)$ is given in the finite plane by the equation $y = 1$. But this is the same line as $\bar{L}_1$, which is parametrized in the finite plane by $(t : 1)$.

In the case of $\overline{L}_2$, we have to solve $ts = 0$, and we get the 2 solutions (up to homotheties) $s = 0, t = 1$ and $s = 1, t = 0$. The first solution gives a point at infinity and the second a point in the finite plane. You can check that the local intersection multiplicity is 1 in both cases.

In the case of $\overline{L}_3$, we have to solve $t^2 = 0$. This gives us the solution $t = 0$, corresponding to the point $(0 : 0 : 1)$. The local intersection multiplicity is 2 and there is no intersection at infinity.

PART II

ELLIPTIC CURVES AND ALGEBRA

Chapter 6

❉ ❉ ◉ ◈ ❉

TRANSITION TO ELLIPTIC CURVES

Road Map

We now can leap in, and begin to describe the objects at the heart of this book: elliptic curves. We begin by looking both backwards, repeating in brief the concepts that we need from Part I, and forwards, to explain where our journey will lead us.

In the first part of this book, we discussed at great length equations in two variables of the form $f(x, y) = 0$, where f is a polynomial of degree $d > 0$ with coefficients in a field K. Such an equation defines a "plane curve" C. For any field K' that contains the field K, we can wonder about the set of all solutions of the equation when the variables are given values from K'. We call this set $C(K')$.

We saw that in order to make the degree of f equal to the number of intersections of any probing line with the curve, we had to do three things:

1. Use an algebraically closed field K'.
2. Work with the projective plane curve $\overline{C}$ defined by $F(x, y, z) = 0$, where F is the homogenization of f.
3. Count intersection points with multiplicities.

From now on, we will usually consider the whole projective plane curve, so we will drop the overline from the notation, and speak of the projective plane curve C.

We also discussed briefly the concept of "singular point" on C.

DEFINITION: We say that $P = (a : b : c)$ is a *singular point* on C if

- It is a point on C, that is, $F(a, b, c) = 0$; and
- There is no algebraically defined tangent line to C at P. In other words, all of the partial derivatives at P vanish:
  $F_x(a, b, c) = F_y(a, b, c) = F_z(a, b, c) = 0$.

We actually only defined singular points "locally." If you are in the coordinate patch defined by $z = 1$, and the curve is defined by $f(x, y) = 0$ in that patch, then $(a : b : 1)$ is a singular point on C if and only $f(a, b) = f_x(a, b) = f_y(a, b) = 0$. Let's show that in this coordinate patch, our old definition is equivalent to the new definition we just gave. (If you are willing to believe this claim, then you can skip the next paragraph.)

Say $f(x, y) = \sum a_{ij} x^i y^j$, a polynomial of degree d. Then $F(x, y, z) = \sum a_{ij} x^i y^j z^{d-i-j}$. So $f(a, b) = 0$ if and only if $F(a, b, 1) = 0$, as we already knew. Now look at the partial derivatives with respect to x: $f_x = \sum i a_{ij} x^{i-1} y^j$ and $F_x = \sum i a_{ij} x^{i-1} y^j z^{d-i-j}$. So $F_x(a, b, 1) = f_x(a, b)$, and if one of these two derivatives equals zero, the other does too. The same kind of thing works for the partial derivatives with respect to y. Finally, let's look at the partial derivative with respect to z. We need to show that if $f_x(a, b) = f_y(a, b) = f(a, b) = 0$, then $F_z(a, b, 1) = 0$. For this we use a trick. Simple algebra shows that because F is homogeneous of degree d, we have the equation $xF_x + yF_y + zF_z = dF$. (This equality is called *Euler's Lemma*. Leonhard Euler was a Swiss mathematician who lived from 1707 to 1783.) Plug in $(a, b, 1)$ for the variables, and you get $aF_x(a, b, 1) + bF_y(a, b, 1) + F_z(a, b, 1) = dF(a, b, 1)$. But we've already seen that $F_x(a, b, 1) = F_y(a, b, 1) = F(a, b, 1) = 0$. So we conclude that $F_z(a, b, 1) = 0$.

As we have mentioned, if L is a probing line to a curve C and if L intersects C at a point P with local intersection multiplicity at least 2, then either L is tangent to C at P, or P is a singular point on C. So if two different lines L_1 and L_2 both intersect C at P with multiplicity at least 2, then P must be a singular point on C. Conversely, if P is a singular point, then *every* line through P has local intersection multiplicity at least 2. If you are up for some algebra, all of these assertions are not too hard to prove. See figure 6.1 to see the difference between a line through a singular point, and a tangent line at a smooth point.

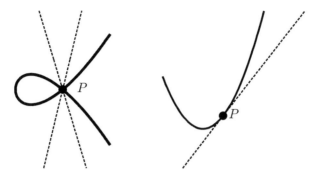

Figure 6.1. Singular vs. nonsingular points with local intersection multiplicity 2

DEFINITION: Let F be a homogeneous polynomial with coefficients in the field K, defining a projective curve C. We say that C is *nonsingular* if for some algebraically closed field K' containing K, every point P in $C(K')$ is nonsingular.

It can be proved that our definition of nonsingular curve doesn't depend on *which* algebraically closed field K' you use. If C is nonsingular and K' is *any* algebraically closed field containing K, then every point P in $C(K')$ will be nonsingular. However, if you change the base field K, the curve defined by F might change its status from nonsingular to singular or vice versa.

We want to use an algebraically closed field in this definition for similar reasons as to why we wanted it when looking at probing lines. For example, consider the curve C defined by $F(x, y, z) = x^2 + y^2 + z^2 = 0$. Then $C(\mathbf{R})$ is the empty set (don't forget that $(0 : 0 : 0)$ is meaningless—it is not the coordinate triple of a point in the projective plane). So to see if C is nonsingular, we really should use a field where it has some points, and look at $C(\mathbf{C})$. Now $F_x(x, y, z) = 2x$, $F_y(x, y, z) = 2y$, and $F_z(x, y, z) = 2z$. All these partial derivatives vanish simultaneously only when $x = y = z = 0$ and that doesn't define a point in projective space (over any field). So C is nonsingular.

For a singular example, consider the curve D defined by $G(x, y, z) = x^2 + y^2 = 0$. Then $G_x(0, 0, 1) = G_y(0, 0, 1) = G_z(0, 0, 1) = G(0, 0, 1) = 0$, and $(0 : 0 : 1)$ is a singular point on D. Therefore, D is not a nonsingular curve.

This example is a little stupid. You can factor $G(x, y, z) = (x - iy)(x + iy)$, so D is the union of two lines M_1 and M_2 (in the complex projective plane). These lines intersect at the origin, and thus the origin must be a singular point. Any probing line L will hit each of M_1 and M_2 at one point, and so L will intersect $M_1 \cup M_2$ with multiplicity 2.

Here is a fun fact: Let F be a homogeneous polynomial, defining a nonsingular projective curve C. Then F must be irreducible. (That is, F cannot be factored into a product of two homogeneous polynomials both of degree at least 1.) The reason for this is similar to what's in the preceding paragraph. If F factors, each factor defines its own curve, and by Bézout's Theorem, these curves must intersect in at least one point. At each point of intersection, any probing curve hits C with multiplicity at least 2, so the intersection must consist of singular points on C.

So for example, let $H(x, y, z) = (x^2 + y^2 + z^2)(x^4 + y^4 + z^4)$. Then $H = 0$ defines a singular projective curve. But $H(\mathbf{R})$ is empty. So you can only see the singular points if you look in $H(\mathbf{C})$. There are examples of this kind of phenomenon even for irreducible polynomials, but they are too complicated to make it worth writing them down here.

Now suppose that F is a homogeneous polynomial with integer coefficients, defining a nonsingular projective curve C. Because we are number theorists who like to see what patterns integers make when added and multiplied together in various ways, we are exceedingly curious about $C(\mathbf{Q})$. This set consists of the solutions to the equation $F = 0$ with rational coordinates. Because Diophantus studied this kind of problem in ancient days, the subject is called *Diophantine equations* or *Diophantine problems*. Diophantine problems make up a huge amount of number theory, and the rest of this book is concerned with them and some of the mathematics that their study has spawned.

In modern times, we have come to a powerful understanding as to how the number of solutions to a Diophantine problem is affected by the degree of the polynomial(s) defining the problem. In the crudest terms, we can ask whether $C(\mathbf{Q})$ is finite (possibly even empty) or infinite. Then we have the following beautiful theorem, the hardest part of which was proved by the German mathematician Gerd Faltings (1954–):

THEOREM 6.1: Let F be a homogeneous polynomial with integer coefficients of degree d, defining a nonsingular projective curve C by the equation $F(x, y, z) = 0$. Then if $d = 1$ or 2, $C(\mathbf{Q})$ is either empty or infinite. If $d \geq 4$, then $C(\mathbf{Q})$ is always finite!

Theorem 6.1 leaves open the question about the size of $C(\mathbf{Q})$ when $d = 3$. It might happen that $C(\mathbf{Q})$ is the empty set, but let's ignore that case. (It is quite important, but we have enough to do already.) So a big question on the frontiers of Diophantine number theory is the following:

> Let F be a homogeneous polynomial with integer coefficients of degree 3, defining a nonsingular projective curve C by the equation $F(x, y, z) = 0$. Assume that $C(\mathbf{Q})$ contains at least one point. Such a curve is called an *elliptic curve*, and usually mathematicians use the letter E for an elliptic curve rather than C. We can ask: How "big" is $E(\mathbf{Q})$?

This is the question we want to discuss in the remainder of this book. We will have to give a concrete meaning to what "how big" means in this context. This will require us to discuss abelian groups in chapter 7. This will help us measure how big $E(\mathbf{Q})$ is from an algebraic point of view. Later, we will describe how to make an analytic function $L(s)$ that encodes some numerical information about $E(\mathbf{F}_p)$ for all prime numbers p; that will require us to talk about generating functions in chapter 11 and analytic functions in chapter 12. For all this, of course, we will have to learn more about the elliptic curves E.

Finally, we will be in a position to explain a conjectural answer to the size question, called the Birch–Swinnerton-Dyer Conjecture. We will conclude the book with a sample application of the conjecture to an old problem in number theory, the "congruent number problem." Along the way we hope to convey some of the excitement involved in making and testing this amazing conjecture, connecting the algebraic and analytic behavior of an elliptic curve.

Chapter 7

ABELIAN GROUPS

Road Map

One of the most crucial things about an elliptic curve E defined by an equation with integer coefficients is that $E(F)$ is an abelian group for every field F over which the equation is nonsingular. So we have to explain what an abelian group is before we begin to dig deeper into elliptic curves. In this chapter, we explore the concept of abelian group, and define a key number associated to many abelian groups: the *rank*, which measures "how big" the group is.

1. How Big Is Infinity?

When we count mathematical objects such as the number of solutions to an equation, the first fundamental distinction we might make is whether the answer is finite or infinite. If it is finite, we can then ask, "How many?"

What if the number is infinite? The German mathematician Georg Cantor (1845–1918) invented a theory of different sizes of infinity. The smallest infinity is called "countably infinite," and it occurs when the infinite set can be put into one-to-one correspondence with the set of all counting numbers 1, 2, 3, ... Can more be said when an infinite set is countably infinite?

The basic question in this book is, "How many rational points are there on a given elliptic curve?" Equivalently, we can ask, "How many rational solutions are there to a given nonsingular cubic equation in two variables with integer coefficients?" The answer is either "finite" or "countably infinite." (The answer cannot be a larger type of infinity than countably infinite, because the set of *all* points in the *xy*-plane with rational coordinates is a countably infinite set.) When the answer is "countably infinite," there *is* something more that can be said.

Because the rational points on an elliptic curve have the structure of a *finitely generated abelian group*, there is a way to measure "how big" that set is, even when it is infinite. Finitely generated abelian groups can have different sizes, as measured by their "rank." The notion of rank is a key ingredient in formulating the BSD Conjecture, and we will now explain what it is.

2. What Is an Abelian Group?

An abelian group is a set with an operation satisfying certain axioms. (Abelian groups are named for Niels Henrik Abel, a Norwegian mathematician who lived from 1802 to 1829.) Let's call our set G. The operation in an abelian group is usually denoted by a plus symbol, but it could be denoted by any symbol, and sometimes is denoted by no symbol at all, like multiplication. We will follow the standard convention when we state the axioms for an abelian group, and use the plus symbol. We will use the notation $G \times G$ for the set of all ordered pairs (a, b) where a and b are in G.

> **DEFINITION**: An abelian group is a set G and an operation $+ : G \times G \to G$ satisfying the axioms:
>
> A1. $(a + b) + c = a + (b + c)$ for any a, b, and c in G.
> A2. $a + b = b + a$ for any a and b in G.
> A3. There is an element in G, denoted 0 and called the *neutral element*, with the property that $0 + a = a + 0 = a$ for any a in G.
> A4. For any element a in G, there is an element denoted $-a$ in G such that $a + (-a) = (-a) + a = 0$ for any a in G.

An example of an abelian group: $G = \mathbf{Z}$, the set of integers, where $+$ is the usual operation of addition of integers.

Here are some more examples of abelian groups. The first two examples give abelian groups because of the axioms of a ring and a field.

Ex. 1: If R is any ring, with its usual addition operation $+$, then $(R, +)$ is an abelian group. This is called "the additive group of R."

Ex. 2: If K is any field, with its usual multiplication operation $\times$, then $(K^\times, \times)$ is an abelian group. (Remember that $K^\times$ is $K - \{0\}$.) This is called "the multiplicative group of K." Usually, we omit the times sign and write ab instead of $a \times b$.

Ex. 3: If n is a positive integer, the modular arithmetic group $(\mathbf{Z}/n\mathbf{Z}, +)$ is an abelian group with n elements.

Ex. 4: Vectors with m coordinates taken from a field K form an abelian group. The elements are ordered m-tuples $(a_1, \ldots, a_m)$ of elements in K and the operation is the usual addition of vectors:
$(a_1, \ldots, a_m) + (b_1, \ldots, b_m) = (a_1 + b_1, \ldots, a_m + b_m)$.

Ex. 5: The one-element group. This is the group that has only one element: 0. (Or if the group is written multiplicatively, then it is just the one element 1.) Of course, $0 + 0 = 0$, and you can check all the axioms without much effort.

We will see some more complicated examples later in this chapter. The main examples of interest in this book will be $E(\mathbf{Q})$ and $E(\mathbf{F}_p)$, where E is an elliptic curve defined by an equation with integer coefficients.

The axioms for a general, not necessarily abelian group, are the same except that A2 is omitted, and a different symbol (not the plus sign) is normally used to denote the group operation.

Using A1 and A2, you can write the sum of any finite number of elements of G in any order and without parentheses, without ambiguity. Also, note that $-a$ may be equal to a. For example, $0 = -0$. Another example of this: In the group $(\mathbf{F}_2, +)$, $1 + 1 = 0$ and therefore $-1 = 1$.

To specify a group, you have to specify two things. One is the set G, and the other is the definition of the operation on any pair of elements of G. That is, you have to tell what $a + b$ is for any pair of elements a and b in G. Strictly speaking, we should speak of the group as a pair $(G, +)$. But often we'll just talk about the group G and you can infer what the operation is from the context.

One way to specify the group operation is to write out the addition table. Another way to specify the operation is by giving a formula for it.

It is easy to prove that there is only one element of G that can be the neutral element, that is, the element 0 that satisfies A3. Similarly, it is easy to prove that for any given element a in G, there is only one element of G that can be the negative of a, that is, the element $-a$ that satisfies A4.

It is customary to write $a - b$ as a shorthand for $a + (-b)$, and it is customary to write na as a shorthand for

$$\overbrace{a + a + \cdots + a}^{n \text{ times}}$$

when n is an integer greater than 1. We also use $1a = a$ and $0a = 0$ on occasion. Also, $(-n)a = -(na)$ if n is positive.

There are occasions when we prefer to use multiplicative notation for the operation in the group. For example, we have the abelian group where G is the multiplicative group of a field. In such cases, we write ab instead of $a + b$, 1 instead of 0, a^{-1} instead of $-a$, and a^n instead of na. This usage is traditional. In fact, it would be very weird to have to write things like $2 + 7 = 14$ in $\mathbf{R}^\times$.

> **DEFINITION**: Let G be an abelian group. A *subgroup* of G is a sub*set* H of G that is itself a group when endowed with the same operation as G.

In particular, if H is a subgroup of G, the neutral element 0 of G must be in H, and H must be closed under addition and subtraction. Conversely, if these things are true, the other axioms of an abelian group hold automatically for H, and H is a subgroup of G.

3. Generations

In the beginning, there was an abelian group G. How did it get there? It had to have the 0-element, but $0 + 0$ is 0, so from 0 you can't make anything new. If G is just the one-element group, there's nothing more to say. But if G is bigger, it has a nonzero element, say x. Since G contains x, it must also contain $-x$. Of course, it might be that $-x = x$, or $-x$ might be a new element. In any case, $-x$ is there in G. By the same token, G must contain $x + x = 2x$, $x + x + x = 3x$, and, in general, nx for any integer n. These elements may not be all different from each other—there could be repetitions, for example if $G = (\mathbf{Z}/n\mathbf{Z}, +)$. In any case, they are all there in G.

Now you can test the axioms, and see that if you define the set H to be all the elements of G of the form nx, for n an integer, then $(H, +)$ is itself a group. We say that H is the subgroup of G "generated" by x and write $H = \langle x \rangle$.

If $H = G$, then *every* element of G is an integral "multiple" of x and we say that G is *generated* by x.[1] We also say that G is a *cyclic group*. In other words, a cyclic group is a group that can be generated by a single element. By definition, this includes the one-element group, which is generated by nothing at all. For example, $(\mathbf{Z}, +)$ is a cyclic group generated by 1. It can also be generated by -1.

Continuing, again choose some element in G. Let's say it's called y. Then we know that G contains all the integral multiples of y. And therefore, since $+$ is the group operation, G also contains every sum of the form $nx + my$ where n and m are any two integers.

Now, if you define the set J to be all the elements of G of the from $nx + my$ for n and m any integers, you can test the axioms and see that $(J, +)$ is itself a group. We say that J is the subgroup of G "generated" by x and y, and write $J = \langle x, y \rangle$. If $G = J$, we say that G is generated by the elements x and y.

It's sort of a mathematical picky thing to do, but we point out that we could take $y = x$. Then J would be H again. *C'est la vie*. We could even take $y = 77x$, and J would be H again. Why would you ever do something redundant like that? Well, you might be in the middle of a proof and not know what y actually is. It might not be a multiple of x, and then again it might be. We have to allow for all eventualities.

In any case, if $G = J$, we say that G is *generated* by the 2 elements x and y. This includes the case where you could get away with just one generator.

Going back to our example of the cyclic group $(\mathbf{Z}, +)$, generated by 1, we see that it can also be generated by two of its elements, for example 1

---

[1] We make the convention that the one-element group is generated by nothing at all. Since 0 has to be there, it doesn't need to be generated. Theological analogies will be left to the reader.

and 77. If we write a list of all abelian groups generated by two elements, we'd have to include **Z**.

A more interesting instance of this comes when we ask: When do two integers a and b generate **Z**? Even though **Z** *can* be generated by just one integer, namely 1, that's not what we are asking about. Can **Z** be generated by 3 and 7? In other words, we are asking: Can any integer be written in the form $3n + 7m$ for some integers n and m? The answer is yes, because the greatest common divisor of 3 and 7 is 1. Thus the simple question of when do two integers a and b generate **Z** leads us into the theory of the greatest common divisor and prime factorization.

Let's generalize this. Suppose S is a subset of an abelian group $(G, +)$. We take the set of everything we get by adding and subtracting various elements of S and call that set $\langle S \rangle$. Remembering that we are only allowed to add or subtract a finite number of elements of G at one time, we can write:

$$\langle S \rangle = \{n_1 x_1 + \cdots + n_k x_k \mid k \geq 0, n_i \in \mathbf{Z}, x_i \in S\}.$$

By convention, if k is the integer 0, we have the empty sum, which by definition is the neutral element 0 in the group.

You can check the axioms and see that $(\langle S \rangle, +)$ is itself a group, called *the subgroup of G generated by S*. If S is finite, say with 10 elements, then you only need k to go up to 10. But if S is infinite, then you might need all possible k's. If $\langle S \rangle = G$, we say that G is *generated* by S.

Human beings are much better at dealing with finite things than infinite things, as a general rule. At least, finite things seem to fit in our imaginations more easily. The class of all abelian groups has two very important subclasses: the class of all finite abelian groups, and the class of all finitely generated abelian groups. A *finite abelian group* G is one where the set G is a finite set. For example, $\mathbf{Z}/n\mathbf{Z}$ is a finite group for any positive integer n. A *finitely generated abelian group* is one where $G = \langle S \rangle$ for some finite subset S. Because $G = \langle G \rangle$, a finite abelian group is always also a finitely generated abelian group. There are many important abelian groups that are not finitely generated, for example $(\mathbf{Q}, +)$, the set of rational numbers with addition as the group operation.

We have lots of nice theorems about finitely generated abelian groups. So if we are curious about some abelian group, and we know it is finitely generated, that knowledge is very helpful in formulating new things we might ask or discover about it.

An example of a *finite* abelian group is the set of vectors $(a_1, \ldots, a_7)$ where the a_i are all in $\mathbf{F}_2$. The group operation is the usual addition of vectors. This group has 2^7 elements.

An example of a *finitely generated* abelian group is the set of vectors $(a_1, \ldots, a_7)$ where the a_i are all in $\mathbf{Z}$. The group operation is again the usual addition of vectors. This group is generated by the set of 7 standard unit vectors $(1, 0, \ldots, 0), \ldots, (0, \ldots, 0, 1)$.

Another example of a nonfinitely generated abelian group is the set of all polynomials in the single variable x with integer coefficients, with group operation addition. Can you see why this is not finitely generated?

EXERCISE: What is the logical difference between saying an abelian group *cannot be finitely generated*, and saying it is *infinitely generated*?

SOLUTION: Every infinite abelian group is infinitely generated, because $G = \langle G \rangle$, but some infinite abelian groups can be finitely generated.

4. Torsion

Suppose G is an abelian group and x is an element of G. It is possible that $x = 0$. If so, we say x has "order 1." If $x \neq 0$, it is possible that $2x = x + x = 0$. If so, we say x has "order 2." In general, suppose there is a positive integer n with the property that $nx = 0$ but $mx \neq 0$ for all positive integers m less than n. If this happens, then we say x has *order n*.

For example, consider the element 2 in the group $(\mathbf{Z}/8\mathbf{Z}, +)$. We have $2 \neq 0, 2 + 2 \neq 0, 2 + 2 + 2 \neq 0$, and $2 + 2 + 2 + 2 = 0$. Therefore, 2 has order 4 in $\mathbf{Z}/8\mathbf{Z}$.

For another example, if $G = (\mathbf{Z}/5\mathbf{Z}, +)$, the orders of the elements are as follows: 0 has order 1, and all other elements have order 5.

EXERCISE: Let $G = (\mathbf{Z}/6\mathbf{Z}, +)$. What are the order of the elements of G?

SOLUTION: The orders of the elements of G are as follows: 0 has order 1, 1 and 5 have order 6, 2 and 4 have order 3, and 3 has order 2.

The orders of the elements of $(\mathbf{Z}/n\mathbf{Z}, +)$ are not hard to figure out, and depend on divisibility properties of integers, as you can see in the previous examples. The orders of elements in $(\mathbf{F}_p^{\times}, \times)$ are much trickier to predict. For example, let $G = (\mathbf{F}_7^{\times}, \times)$. What is the order of 2? You have to start multiplying (which is the operation in this group): $2 = 2^1 \neq 1$, $4 = 2^2 \neq 1$, but $8 = 2^3 = 1$ in G, so the order of 2 in $(\mathbf{F}_7^{\times}, \times)$ is 3.

DEFINITION: Let G be an abelian group and x an element of G. We say x is a *torsion element* of G if x has order n for some integer $n \geq 1$. If x is not a torsion element of G, we say x has *infinite order*.

Some handy informal terminology:

DEFINITION: Let G be an abelian group and x an element of G. Let k be any integer. If $kx = 0$, we say k *kills* x.

Some people object to this bloodthirsty language, but it is very convenient. Notice that if any nonzero integer kills x, then x has finite order. Why? Suppose $kx = 0$, $k \neq 0$. We can't say that the order of x is k, because k may not be positive. Even if k is positive, k may not be the *least* positive integer that kills x. But if k is negative, $(-k)x = -(kx) = -0 = 0$ so x is killed by some positive integer. Then there is the least positive integer n that kills x, and x will have order n. Note that if no positive integer kills x, then x has infinite order by definition.

For example, take the group $(\mathbf{Z}, +)$. Every element has infinite order except 0, so that 0 is the only torsion element in $\mathbf{Z}$. Another example: In the group $(\mathbf{R}^{\times}, \times)$, the only torsion elements are ± 1. All the other elements have infinite order. This is because if x is a real number, and $x^n = 1$ for some $n > 1$, then $x = \pm 1$.

EXERCISE: Let $G = (\mathbf{C}^{\times}, \times)$. What are the orders of the elements of G? Does G have torsion elements other than 1? How many?

EXERCISE: Let $(G, +)$ be any abelian group. Let H be the subset of G consisting of all the torsion elements of G. In other words, H consists of all elements of finite order in G. Check that H is itself a group, under the same operation $+$ as G. It is called "the torsion subgroup of G."

SOLUTION: First, we have to check that $+$ is an operation on H. In other words, if x and y are both in H, we have to check that $x + y$ is also in H. Well, if x is in H, then it has some finite order, say n. So $nx = 0$. Similarly, if y is in H, then it has some finite order, say m. Therefore $(nm)(x + y) = nmx + nmy = m(nx) + n(my) = m0 + n0 = 0 + 0 = 0$. So $x + y$ has finite order and is in H.

Next we have to check the axioms. A1 and A2 are true for H because H is a subset of G. A3 is true for H because 0 has finite order (namely it is order 1) and so is in H. Finally, A4 is true for H because if a is in H, then a killed by some positive integer k, that is, $ka = 0$, and therefore the equation $k(-a) = -(ka) = -0 = 0$ shows that $-a$ also has finite order. Hence, $-a$ is in H.

Now we come to a very important theorem:

THEOREM 7.1: If G is a finitely generated abelian group, each of its subgroups is also finitely generated. In particular, its torsion subgroup is finitely generated, and therefore finite.

We will not prove this theorem in this book. It sounds plausible, but in fact is a bit tricky to prove.

To summarize, if G is a finitely generated group, then its torsion subgroup H is a finite group. And any finite group equals its own torsion subgroup. For example, the group $\mathbf{Z}/1000\mathbf{Z}$ is finitely generated (in fact, it is cyclic) and it is its own torsion subgroup.

5. Pulling Rank

In this section, we'll be discussing an abelian group $(G, +)$, which is assumed to be finitely generated. We won't keep repeating the assumption that G is finitely generated. If G is equal to its torsion subgroup, it is finite. With a sufficiently lofty view, we might even say it is "negligible."

But what if G is not finite, and therefore not equal to its torsion subgroup? Is there any sense in which we can say how "big" G is, beyond saying it has infinitely many elements? Yes, there is an important sense in which we can do this.

To warm up for this, consider the question: Which is "bigger": a line, a plane or all of space? That is to say, which is bigger: $\mathbf{R}^1$, $\mathbf{R}^2$, or $\mathbf{R}^3$? These are all infinite sets, and there are one-to-one correspondences between any two of them, so from the point of view of set theory, they are all the same "size."

But surely we feel a plane is "bigger" than a line, and that space is "bigger" than a plane? Why? In a line, there is only one direction to go forward or back in, but in a plane, there are infinitely many directions. OK, but in space there are infinitely many directions also. Why do we feel that space is "bigger" than a plane? In a plane, there are only two "independent" directions. But in space, there are three.

Along the same lines, we can ask how many "independent directions" there are in a finitely generated abelian group. The rigorous way of doing this is to speak in terms of generators for the group. The more "independent directions" there are, the more elements we need to generate the group.

> **DEFINITION**: Let G be a finitely generated abelian group and let H be its torsion subgroup (which is necessarily finite). The smallest integer r so that G can be generated by r elements along with all of the elements of H is called the *rank* of G.

If G is a finite group, it equals its torsion subgroup, so $r = 0$. So the rank of a finite group is 0. If G is infinite, it must have rank r at least 1. (Remember we are assuming through this section that G is finitely generated.)

EXAMPLES:

Ex. 1. $\mathbf{Z}/1000000\mathbf{Z}$ is a finite group, so it has rank 0.

Ex. 2. $\mathbf{Z}$ is infinite. It can be generated by 1 element (namely

the element 1) and not by fewer elements, so it has
rank $r = 1$.

Ex. 3. $\mathbf{Z}^2$ by definition is the group of ordered pairs of
integers under addition. It can be generated by 2
elements (e.g., $(1, 0)$ and $(0, 1)$) and not by fewer
elements, so it has rank $r = 2$.

Ex. 4. $\mathbf{Z}^9$ by definition is the group of ordered 9-tuples of
integers under addition. It can be generated by 9
elements (e.g.,
$(1, 0, 0, 0, 0, 0, 0, 0, 0), \ldots, (0, 0, 0, 0, 0, 0, 0, 0, 1))$ and
not by fewer elements, so it has rank $r = 9$.

In Ex. 2–4, the torsion subgroup consists only of the neutral
element in the group.

Ex. 5. G is the funny group defined like this: The elements of
G are ordered triples (a, b, c) where a is an integer, b is
an element of $\mathbf{Z}/10\mathbf{Z}$, and c is an element of $\mathbf{Z}/11\mathbf{Z}$.
The group law is addition of triples:
$(a, b, c) + (d, e, f) = (a + d, b + e, c + f)$. Notice that
the meaning of the plus sign is subtly different in each
place, because the plus sign can refer to three different
groups with three different definitions of addition.
Now you can check that the torsion subgroup H of G
consists of all triples (a, b, c) with $a = 0$. Then the
single element $(1, 0, 0)$, together with all of the torsion
elements, generate G, because any element (a, b, c) in G
can be written as
$a(1, 0, 0) + (0, b, c) = a(1, 0, 0) +$ a torsion element. So
the rank of G is 1.

Appendix: An Interesting Example of Rank and Torsion

The main example we consider in this book is $E(\mathbf{Q})$, where E is a smooth
projective curve defined by a homogeneous integral polynomial of degree
3. To finish this chapter, we will look at a simpler example that also comes
from number theory. This example arose historically from the problem of

finding all integral solutions to equations of the form $x^2 - Ny^2 = 1$, where N is a positive integer.

We start with an irreducible polynomial $f(x)$ of degree $n > 1$ and with integer coefficients. "Irreducible" means that you can't factor $f(x) = g(x)h(x)$, where g and h are both nonconstant polynomials with integer coefficients. We also assume that the leading coefficient of f is 1. So $f(x) = x^n + a_{n-1}x^{n-1} + \cdots + a_1x + a_0$, where $a_0, a_1, \ldots, a_{n-1}$ are all integers. For instance, we could use $f(x) = x^2 - 7$.

Let θ be a root of $f(x)$ in **C**. (Since **C** is algebraically closed, there certainly is a root like this—in fact, $f(x)$ will have n roots, all of them contained in **C**.) We form the group $(U, \times)$ as follows: U will be the set of complex numbers of the form

$$x = b_0 + b_1\theta + b_2\theta^2 + \cdots + b_{n-1}\theta^{n-1}$$

where $b_0, b_1, \ldots, b_{n-1}$ are all integers *and* there exists another such number

$$y = c_0 + c_1\theta + c_2\theta^2 + \cdots + c_{n-1}\theta^{n-1}$$

with $c_0, c_1, \ldots, c_{n-1}$ all integers such that $xy = 1$. The group law on U is multiplication.

No matter what f and what θ you start with, it turns out U is a finitely generated abelian group. What is the rank of U? The answer depends on f, of course, and is contained in what is called *Dirichlet's Unit Theorem*. We won't attempt to explain the proof, but we can explain what the theorem says.

Here's the answer: You look at all the roots of $f(x)$. There will be n of them. Some of the roots will be real numbers, and some will be nonreal, complex numbers. Suppose there are r real roots and s pairs of nonreal roots. (The nonreal roots have to come in conjugate pairs. If θ is a nonreal root of $f(x)$, then its complex conjugate $\bar{\theta}$ is also a root. Why? $f(\bar{\theta}) = \overline{f(\theta)}$ because complex conjugation respects addition and multiplication, and the coefficients of f are all real numbers. Then $\overline{f(\theta)} = f(\bar{\theta}) = \bar{0} = 0$.) Dirichlet's Unit Theorem says that the rank of U is $r + s - 1$.

This theorem, which goes back to the first part of the nineteenth century, is not that hard to prove, but not that easy either. In some ways, it is a prototype of the type of question we will be asking later about the rational points on an elliptic curve. These questions all involve the rank of abelian groups that are constructed out of interesting number-theoretical situations. In the case of these "unit groups," Dirichlet provides us with a complete answer, depending only on the roots of $f(x)$. In the case of elliptic curves, there is an answer that depends on properties of a much more complicated object, the L-function. As of today, this answer in the elliptic curve case is only a conjecture—the BSD Conjecture whose elucidation is our goal in this book.

If you would prefer to see an example with the group law written additively, we could form the group V consisting of the set of real numbers $\log(|u|)$ where u runs through all elements of U. This V is a group under the operation of addition of real numbers, and V has the same rank as U: $r + s - 1$. In fact, taking logarithms is part of the proof of Dirichlet's Unit Theorem.

For an interesting set of examples of Dirichlet's Unit Theorem in action, you could take any odd prime number p and consider the polynomial $f_p(x) = x^{p-1} + x^{p-2} + \cdots + x + 1$. The roots of $f_p(x)$ are all the primitive p-th roots of unity, that is, all the solutions of the equation $x^p = 1$ except for the solution $x = 1$, because $f_p(x) = (x^p - 1)/(x - 1)$. These roots of unity come in $(p - 1)/2$ pairs of nonreal numbers. So $s = (p - 1)/2$ and the rank of U in this case $(p - 1)/2 - 1 = (p - 3)/2$. However, knowing the rank of U is far less useful than knowing how to construct all of the elements of U, which is a much deeper problem in the theory.

The study of these groups U and related groups packs in a lot of number theory, most of which we cannot explain here. But we can do the example where $f(x) = x^2 - N$ (N a nonsquare integer) in more detail, and we'll see that Dirichlet's Unit Theorem in this case tells us something very powerful about the famous Diophantine equation $x^2 - Ny^2 = 1$ that we mentioned at the beginning of this section.

Choose your favorite nonsquare integer N and set $f(x) = x^2 - N$. The roots of $f(x)$ are $\pm\sqrt{N}$. If N is positive, we get two real roots, so $r = 2$, $s = 0$, and the rank of U is $2 + 0 - 1 = 1$ by Dirichlet's Unit Theorem. If N is negative, we get one pair of nonreal roots, so $r = 0$, $s = 1$, and the rank of U is $0 + 1 - 1 = 0$. This computation is very interesting: When the rank is 0, U must equal its own torsion subgroup. So in this case, U is finite. More careful analysis can determine that U consists of "roots of unity," and will have either 2, 4, or 6 elements.

Let's look at the case where $N > 0$ further. One of the roots of $f(x) = 0$ is $\theta = \sqrt{N}$. Then U is the set of all real numbers of the form $a + b\sqrt{N}$ where a and b are integers *and* there exists integers c and d such that $(a + b\sqrt{N})(c + d\sqrt{N}) = 1$.

We define the *norm* of $t = a + b\sqrt{N}$ to be the real number $\mathcal{N}(t) = a^2 - Nb^2$. (You will see this idea of the norm again later, when we discuss singular cubic curves in chapter 9.) If a and b are integers, then of course $\mathcal{N}(t)$ is an integer too. For example, $\mathcal{N}(2 + 3\sqrt{5}) = 2^2 - 5 \cdot 3^2 = -41$. Another example: $\mathcal{N}(1) = \mathcal{N}(1 + 0\sqrt{N}) = 1^2 - N \cdot 0^2 = 1$.

We claim that U consists of all $t = a + b\sqrt{N}$ such that $\mathcal{N}(t) = \pm 1$. Why should something like this be true? It's because the norm is multiplicative.

EXERCISE: Prove that if $t = a + b\sqrt{N}$ and $t' = a' + b'\sqrt{N}$ then $\mathcal{N}(tt') = \mathcal{N}(t)\mathcal{N}(t')$.

SOLUTION: This is sheer algebra. On the one hand,
$\mathcal{N}(tt') = \mathcal{N}((a + b\sqrt{N})(a' + b'\sqrt{N})) = \mathcal{N}(aa' + bb'N + (ab' + a'b)\sqrt{N}) = (aa' + bb'N)^2 - N(ab' + a'b)^2 =$
$(aa')^2 + 2aa'bb'N + (bb'N)^2 - N((ab')^2 + 2aa'bb' + (a'b)^2) =$
$(aa')^2 + (bb'N)^2 - N((ab')^2 + (a'b)^2)$ because the terms $\pm 2aa'bb'N$ cancel.

On the other hand, $\mathcal{N}(t)\mathcal{N}(t') = (a^2 - Nb^2)(a'^2 - Nb'^2)$. When you multiply this out, you get the same result that we just computed for $\mathcal{N}(tt')$.

We know that U consists of $t = a + b\sqrt{N}$ such that there exists $u = c + d\sqrt{N}$ with a, b, c, and d all integers, and $tu = 1$. But if $tu = 1$, then $\mathcal{N}(t)\mathcal{N}(u) = \mathcal{N}(1) = 1$. But $\mathcal{N}(t)$ and $\mathcal{N}(u)$ are both integers. So either $\mathcal{N}(t) = \mathcal{N}(u) = 1$ or $\mathcal{N}(t) = \mathcal{N}(u) = -1$.

Conversely, suppose you have $t = a + b\sqrt{N}$ with $\mathcal{N}(t) = \pm 1$. Then we claim you can take $u = c + d\sqrt{N} = (a - b\sqrt{N})/\mathcal{N}(t)$ and get $tu = 1$. (Note that c and d are still integers because we are dividing the integers a and b only by 1 or -1.)

Let's see:

$$tu = (a + b\sqrt{N})(a - b\sqrt{N})/\mathcal{N}(t) = (a^2 - Nb^2)/\mathcal{N}(t) = \mathcal{N}(t)/\mathcal{N}(t) = 1.$$

Figure 7.1. A 7-gon inscribed in a circle

Thus, the norm function allows us to simplify the study of U. Instead of fishing around for whether a u exists such that $tu = 1$, we can directly compute the norm $\mathcal{N}(t)$ and see whether $\mathcal{N}(t) = \pm 1$. In the case where we start with a more complicated polynomial for $f(x)$, there is also a norm function, but we won't go into that.

To summarize so far: Given a nonsquare positive integer N, we form the set of real numbers $U = \{a + b\sqrt{N} \mid a^2 - Nb^2 = \pm 1\}$. Ordinary multiplication of real numbers defines an operation on U, and U becomes an abelian group of rank 1. To see more explicitly what this means, let's ask about the torsion subgroup of U.

Suppose v in U has finite order. This means that $v^k = 1$ for some positive integer k. But this means v is a k-th root of unity. Remember what they look like in the complex plane? All the k-th roots of unity in the complex plane form the vertices of a regular k-gon inscribed in the circle of radius 1, as in figure 7.1.

Now every element of U is a *real* number. So this v has to be a real root of unity, and the only possibilities are 1 and -1. We conclude that the torsion subgroup of U is the two-element group $\{1, -1\}$.

The fact that the rank of U is 1 means that U can be generated by -1 and one additional element. What this boils down to saying (we skip a few small steps) is that the equation

$$a^2 - Nb^2 = 1 \tag{7.2}$$

has a solution (a_1, b_1), with $a_1 > 0$ and $b_1 > 0$, such that all other such solutions may be found from this one. How? You write $t_1 = a_1 + b_1\sqrt{N}$. Define the number $t_m = t_1^m$ for any integer m, positive or negative. Write out $t_m = a_m + b_m\sqrt{N}$. Then (a_m, b_m) is again a solution of (7.2), and all solutions with positive a_m and b_m are obtained this way.

The fact that there is a single "fundamental" solution like this is very nonobvious. It is a consequence of the fact that the rank of U is 1. Just knowing there is a fundamental solution is weaker than being able to find it. The theory of continued fractions enables you to find the fundamental solution. For example, if $N = 7$, the fundamental solution of $a^2 - 7b^2 = 1$ is given by $a = 8$ and $b = 3$. You can read about continued fractions and their application to equation (7.2) in many elementary number theory books, including (Davenport, 2008).

Equation (7.2) is usually called "Pell's equation." It has a very long history, which is summarized in (Weil, 2007). Apparently, the English mathematician John Pell (1611–85) didn't have much to do with it, but got his name attached to the equation because of a mistake by Euler. Is this real fame?

Chapter 8

. . . . .

NONSINGULAR CUBIC EQUATIONS

Road Map

Yet again, we see that the interplay between algebra and geometry leads to fascinating mathematics. In this case, we use geometry to construct the abelian group operation on the points of an elliptic curve. Then we derive algebraic formulas for this operation.

1. The Group Law

How to turn a nonsingular cubic curve into an abelian group is not that hard to describe geometrically. The hard part, which we will not perform fully, is verifying that all of the axioms described in chapter 7 are satisfied.

By now, we have arranged our definitions so that any line intersects the curve defined by a cubic equation in 3 points, provided that we use the definition of "intersect" that we have constructed. Our goal in this chapter is to make use of some consequences of that fact.

We begin with a nonsingular cubic curve E, defined over a field K. In particular, E is defined by an equation whose coefficients are in K. We assume that E has at least one point with coordinates in K. (You'll have to wait until chapter 9 to learn what remains the same and what changes when we start instead with a singular cubic equation.) For example, by definition, an elliptic curve defined over $\mathbf{Q}$ is assumed to possess at least one point with rational projective coordinates. (We call such a point a "rational point" on the curve.) This is a less innocuous assumption than it sounds. If you have a homogeneous polynomial in three variables with rational coefficients, defining the projective curve C, it can be difficult to tell whether C has any rational points at all. There is no known algorithm for doing this.[1]

> **DEFINITION**: Let K be a field. A nonsingular cubic curve over K containing at least one point with K-coefficients is called an *elliptic curve* over K. We call the given point $\mathcal{O}$.

Our first order of business is to explain why the points on an elliptic curve form a group. Let's start with two points P and Q, that are both elements of $E(K)$. In this section, we describe the recipe geometrically, and let the algebra wait for a few pages.

Step 1: Let L be the line connecting P and Q. If $P = Q$, this means that L is the tangent line at P. (The requirement that there is a unique tangent line at every point on E is one reason why we restrict ourselves to nonsingular cubics.) The line L intersects E at a third point R. See figure 8.1.

Step 2: Draw a line L' from R to $\mathcal{O}$. (If $R = \mathcal{O}$, then draw the tangent line at $\mathcal{O}$.) The line L' again intersects E at a third point. This third point of intersection is defined to be $P + Q$. See figure 8.2.

Notice that because we give the recipe in terms of the line connecting P and Q, it is immediately obvious that $P + Q = Q + P$; the order in which we give the points is irrelevant. So if this $+$ operation does define a group, it will be an abelian group.

There is no reason in the world why you should believe without further explanation that this recipe defines a group at all. We need to talk about the identity element, the inverse of an element, and the associative property (i.e., properties A3, A4, and A1 on page 101). It is also not immediately obvious that if the coordinates of P and Q are both in the field K, then the coordinates of $P + Q$ are also in the field K.

---

[1] The existence of cubic projective curves without any rational points but which have points over $\mathbf{F}_p$ for every prime p is closely connected to the existence of nontrivial Tate–Shafarevich groups, named after American mathematician John Tate (1925–) and Russian mathematician Igor Shafarevich (1923–). If you have the technical background, you can read about these groups in (Silverman, 2009). They are mentioned again when we state the strong form of the BSD Conjecture on p. 237.

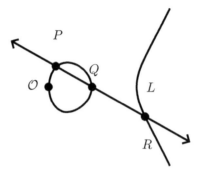

Figure 8.1. Step 1

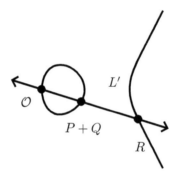

Figure 8.2. Step 2

First, we'll show that the identity element is the point $\mathcal{O}$. We need to verify that if P is any point on E, then $P + \mathcal{O} = P$. Let's follow the recipe. We draw line L connecting P and $\mathcal{O}$, and it intersects the curve in a third point R. Now we draw a line from R to $\mathcal{O}$. It intersects the curve at a third point, and *that third point must be P*, because we know that P, $\mathcal{O}$, and R are colinear, and that any probing line L hits E at 3 and only 3 points (counted with multiplicity, of course). So our recipe says that $P + \mathcal{O} = P$. See figure 8.3.

What about the inverse of P, which incidentally is always written as $-P$? Draw the tangent line at $\mathcal{O}$, and the tangent will intersect E at a third point $\mathcal{O}'$. (If $\mathcal{O}$ is an inflection point, as it will be when we make some choices below, then $\mathcal{O}' = \mathcal{O}$. With this choice, the description of the group law in terms of colinearity given in the Prologue is correct.) Draw the line from $\mathcal{O}'$ to P, and that line intersects E in a third point. That third point is $-P$. To summarize: Our construction says that P, $\mathcal{O}'$, and $-P$ are colinear.

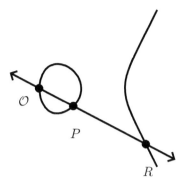

Figure 8.3. Verification of the identity element

How do we check that this point $-P$ really is the inverse of P? We follow the geometric recipe to add P and $-P$. First, we connect P and $-P$, and this line intersects E at $\mathcal{O}'$. Then we perform the second step in our recipe, and connect $\mathcal{O}'$ to $\mathcal{O}$. The third point of intersection is $P + (-P)$. But the third point of intersection is again $\mathcal{O}$, because the line through $\mathcal{O}$ and $\mathcal{O}'$ is tangent at $\mathcal{O}$.

Finally, what about the associative property? That's just too hard to do geometrically in this book. (Associativity turns out to be implied by a classical result in projective algebraic geometry.) We'll discuss below how to describe the $+$ operation algebraically, and then (if you have enough perseverance) you can do a lot of algebra and verify that $(P_1 + P_2) + P_3 = P_1 + (P_2 + P_3)$.

2. Transformations

Let E be an elliptic curve. Following (Silverman and Tate, 1992), eventually you find that, by choosing coordinates in a clever way, the projective equation for E may be assumed to take the form:

$$y^2 z + a_1 xyz + a_3 yz^2 = x^3 + a_2 x^2 z + a_4 xz^2 + a_6 z^3. \qquad (8.1)$$

The point that we called $\mathcal{O}$ now has projective coordinates $(0 : 1 : 0)$, and moreover $\mathcal{O}$ is the *only* point on the intersection of E and the projective line $z = 0$. That last fact means that the local intersection multiplicity of the line $z = 0$ and E at $\mathcal{O}$ must be 3 (i.e., $\mathcal{O}$ is an inflection point). Because the line $z = 0$ is the line at infinity, this means that $\mathcal{O}$ is the only point at infinity on E, in these coordinates.

EXERCISE: Verify that $(0 : 1 : 0)$ does indeed solve equation (8.1), and that $\mathcal{O}$ is the only point on the intersection of E and the line $z = 0$.

Because there is only one point on E at infinity, we can set $z = 1$ in equation (8.1), work with the equation

$$y^2 + a_1 xy + a_3 y = x^3 + a_2 x^2 + a_4 x + a_6 \tag{8.2}$$

and say that the points on $E(K)$ form the set

$$\left\{\; (x, y) \;\middle|\; x, y \in K, \quad y^2 + a_1 xy + a_3 y \right.$$
$$\left. = x^3 + a_2 x^2 + a_4 x + a_6 \;\right\} \; \bigcup \; \left\{\mathcal{O}\right\}.$$

In fact, even equation (8.2) can be simplified. Assuming that the characteristic of K is neither 2 nor 3, it is possible to make another change of variables, and rewrite the equation in the form

$$E: \qquad y^2 = x^3 + Ax + B. \tag{8.3}$$

We will mostly work with this simplest form of the equation for E, but we will also give a few formulas using equation (8.2).

There are at least two questions that should be on your mind:

1. What's the point of the strange numbering system in equation (8.2)? Why is there no a_5?
    ANSWER: Though (8.2) is obviously not homogeneous, there is a sense in which it can be thought of as "nearly" homogeneous. If we count x with "weight" 2 and y with "weight" 3, and a_k with "weight" k, then every term in the equation has weight 6. A systematic way to think of this is that if we replace x by ξ^2, and replace y by η^3, and replace a_k by α_k^k (for $k = 1, 2, 3, 4,$ and 6), then the result is a homogeneous equation of degree 6. We will not pursue the consequences of this observation, but that's the explanation for the strange (but traditional) numbering on the constants a_k, as well as the constants b_k and c_k defined below.
2. Something more fundamental should have occurred to you. Why did we need to make our equations so complicated? Why can't we just work with an equation of the form $y = x^3 + Ax + B$ instead of $y^2 = x^3 + Ax + B$?

ANSWER: Suppose that we start with an equation of this form. It appears to meet our criteria (nonsingular cubic equation containing a point), but in fact it is singular! The problem is that the singular point is hidden away at infinity, where we do not see it right away.

Take the homogeneous form of the equation: $yz^2 = x^3 + Axz^2 + Bz^3$. You can check that the projective point $(0 : 1 : 0)$ solves this equation, and moreover you can verify that if $F(x, y, z) = yz^2 - x^3 - Axz^2 - Bz^3$, then $F_x(0 : 1 : 0) = F_y(0 : 1 : 0) = F_z(0 : 1 : 0) = 0$. In other words, there is a singularity at $(0 : 1 : 0)$, so the curve is not nonsingular. In fact, this curve is an example of what we will call "additive reduction" in chapter 9.

3. The Discriminant

Typically, we will start with a cubic equation either in the form of equation (8.2) or (8.3), where the coefficients are integers. We need to ensure that the curve defined by the cubic is not singular, and the recipe is a bit complex (particularly if we start with equation (8.2)) and extremely well-suited to computer programming.

We define

$$b_2 = a_1^2 + 4a_2$$

$$b_4 = a_1 a_3 + 2a_4$$

$$b_6 = a_3^2 + 4a_6$$

$$b_8 = b_2 a_6 - a_1 a_3 a_4 + a_2 a_3^2 - a_4^2$$

$$c_4 = b_2^2 - 24b_4$$

$$c_6 = -b_2^3 + 36b_2 b_4 - 216b_6$$

$$\Delta_E = -b_2^2 b_8 - 8b_4^3 - 27b_6^2 + 9b_2 b_4 b_6 \tag{8.4}$$

$$j_E = \frac{c_4^3}{\Delta_E}$$

Then a theorem tells us that the curve E, as defined by equation (8.1), is singular if and only if $\Delta_E = 0$ in the field that E is defined over. The constant Δ_E is called the *discriminant* of the curve E. These formulas come from a standard computation in algebraic geometry that determines where an algebraic variety is singular.

Why did we bother defining c_4, c_6, and j_E? An alternate formula for Δ_E is $\Delta_E = \frac{c_4^3 - c_6^2}{1728}$. As for j_E, we in fact will not be using it in our exposition, but it is an important number attached to the elliptic curve, called the *j-invariant* of E.

> **EXERCISE**: Suppose that the equation for E is given in the form of equation (8.3). Show that $\Delta_E = -16(4A^3 + 27B^2)$.

> **SOLUTION**: Using equation (8.3) amounts to setting $a_1 = a_2 = a_3 = 0$, $a_4 = A$, and $a_6 = B$. We can then compute that $b_2 = 0$, $b_4 = 2a_4 = 2A$, and $b_6 = 4a_6 = 4B$. Finally,
> $$\Delta_E = -8(2A)^3 - 27(4B)^2 = -16(4A^3 + 27B^2).$$

Let's give an example of all of this algebra before we get to the main point of this chapter. Suppose that E is defined by the equation $y^2 = x^3 + 1$. Then $\Delta_E = -16 \cdot 27$. If K is any field of characteristic 2 or 3, then $E(K)$ is singular, because $-16 \cdot 27 = 0$ in such a field, and we banish it to the next chapter. Otherwise, if K is not a field with characteristic 2 or 3, then $E(K)$ is an elliptic curve, and we keep it in this chapter.

In general, if E is an elliptic curve given by equation (8.3), so that $\Delta_E = -16(4A^3 + 27B^2)$, and p is any prime number that does not divide Δ_E, then the curve $E(\mathbf{F}_p)$ is nonsingular. In fact, if K is *any* field of characteristic p for p a prime number that does not divide Δ_E, then the curve $E(K)$ will be nonsingular. We will take advantage of this observation in a moment.

4. Algebraic Details of the Group Law

We have no choice but to wade into algebraic thickets now, and de-scribe how to add two points P and Q on an elliptic curve E given by equation (8.2). We saw above that $P + \mathcal{O} = P$, so we may as well assume that both P and Q solve equation (8.2).

Suppose that $P = (x_1, y_1)$ and $Q = (x_2, y_2)$. Let λ be the slope of the line connecting P and Q, which we called L above. In other words,

- If $x_1 \neq x_2$, then $\lambda = \frac{y_2 - y_1}{x_2 - x_1}$.
- If $x_1 = x_2$ and $y_1 = y_2$, then let λ be the slope of the tangent line at P. Concretely, if the equation for E is given by equation (8.3), then $\lambda = \frac{3x_1^2 + A}{2y_1}$. If the equation for E is given in the more complicated form (8.2), then $\lambda = \frac{3x_1^2 + 2a_2x + a_4 - a_1y_1}{2y_1 + a_1x_1 + a_3}$.
- If $x_1 = x_2$ and $y_1 \neq y_2$, then we will see that $P + Q = \mathcal{O}$.

We need to find the third point of intersection of this line with E. We called that point R above. Suppose that R has coordinates (x_3, y_3), and suppose that the equation for L is $y = \lambda(x - x_1) + y_1$. Substitute the equation for L into the formula for E, and we get a cubic equation involving x. Two of the solutions of that equation are x_1 and x_2. We seek the third solution x_3.

Now, we rely on a helpful fact from algebra. The three solutions of the equation $x^3 + ax^2 + bx + c = 0$ add up to $-a$. Why? Suppose that the three solutions are x_1, x_2, and x_3. Then we know that $x^3 + ax^2 + bx + c = (x - x_1)(x - x_2)(x - x_3)$, and multiplying the right-hand side and looking at the coefficient of x^2 tells us that $-a = x_1 + x_2 + x_3$. In our case, we prefer to write this as $x_3 = -a - x_1 - x_2$.

So we need to write out the cubic equation. First suppose that E is given by the simpler equation (8.3). Substitute $y = \lambda(x - x_1) + y_1$ into (8.3), and we get

$$x^3 - (\lambda(x - x_1) + y_1)^2 + Ax + B = 0.$$

We see that the coefficient of x^2 is $-\lambda^2$, and therefore $x_3 = \lambda^2 - x_1 - x_2$. Fortunately, it's easy to find that $y_3 = \lambda(x_3 - x_1) + y_1$.

The second step of our recipe says that we need to connect $\mathcal{O}$ and R with a line, and where that line intersects E will be the point $P + Q$. The line connecting $\mathcal{O}$ and R is vertical and easy to describe in projective coordinates: it is $x = x_3z$. The third point of intersection of this line and E has coordinates $P + Q = (x_3, y_4)$. If we look at (8.3), we see that y_3 and y_4 must be the two solutions to the equation $y^2 = x_3^3 + Ax_3 + B$, so $y_4 = -y_3 = -(\lambda(x_3 - x_1) + y_1)$.

EXCERCISE (FOR ALGEBRA-LOVERS): Suppose instead that the equation for E is given by equation (8.2). Determine the formulas for x_3, y_3, and y_4.

SOLUTION: We already did some of the work, because above we have the formula for λ, which in turn gives us the formula for the line L. If we substitute $y = \lambda(x - x_1) + y_1$ into (8.2), we get

$$(\lambda(x - x_1) + y_1)^2 + a_1 x(\lambda(x - x_1) + y_1) + a_3(\lambda(x - x_1) + y_1)$$
$$= x^3 + a_2 x^2 + a_4 x + a_6.$$

Bring all of the terms to one side of the equation, and the coefficient of x^2 becomes $a_2 - \lambda a_1 - \lambda^2$. Therefore, $x_1 + x_2 + x_3 = \lambda^2 + \lambda a_1 - a_2$, so $x_3 = \lambda^2 + a_1\lambda - a_2 - x_1 - x_2$. If $R = (x_3, y_3)$, then $y_3 = \lambda(x_3 - x_1) + y_1$, as before.

Again, the line connecting R and $\mathcal{O}$ has the projective equation $x = x_3 z$, and again that means that y_3 and y_4 are the two roots of the equation

$$y^2 + a_1 x_3 y + a_3 y = x_3^3 + a_2 x_3^2 + a_4 x_3 + a_6.$$

We see that $y_3 + y_4 = -a_1 x_3 - a_3$, and so $y_4 = -a_1 x_3 - a_3 - y_3 = -a_1 x_3 - a_3 - \lambda(x_3 - x_1) - y_1$.

EXCERCISE: Show that if $x_1 = x_2$ and $y_1 \neq y_2$, then $P + Q = \mathcal{O}$.

SOLUTION: In this case, the line connecting P and Q is vertical and has the equation $x = x_1$, which in projective coordinates has the form $x = x_1 z$. The third point of intersection of the line L with E is therefore at $\mathcal{O}$. Step 2 of our recipe now says to connect $\mathcal{O}$ with $\mathcal{O}$, and see where the resulting line intersects E. We checked above that the tangent line at $\mathcal{O}$ does not intersect E at any other point, because the tangent line in fact intersects E with multiplicity 3. So line L' touches E only at $\mathcal{O}$, meaning that $P + Q = \mathcal{O}$.

There is a crucial consequence of these computations (which, incidentally, are quite simple to program on a computer, however tedious they

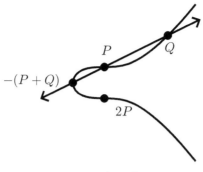

Figure 8.4. $y^2 = x^3 + 1$

are to perform by hand): If x_1, x_2, y_1, and y_2 are all elements of the field K, then the coordinates of $P + Q$ will also be elements of the field K. So the set of points $E(K)$ is closed under this group operation.

We have not explained how to show that this group operation is associative. If you have a lot of time, you can program the above computations, and verify with a computer algebra program that associativity holds. There is also a more abstract way to prove associativity, which you can find in (Silverman and Tate, 1992), for example.

5. Numerical Examples

We should do a few numerical examples. Let's take the elliptic curve given by the equation $y^2 = x^3 + 1$. Suppose first that we are working over the field $\mathbf{Q}$. Let $P = (0, 1)$ and $Q = (2, 3)$ be our two points, and let's compute $P + Q$. We start by computing that $\lambda = 1$, and in fact the line connecting P and Q has equation $y = x + 1$. Then $x_3 = \lambda^2 - x_1 - x_2 = 1^2 - 0 - 2 = -1$, and then the y-coordinate of $P + Q$ is given by $-(\lambda(x_3 - x_1) + y_1) = -(-1 + 1) = 0$. So $P + Q = (-1, 0)$. See figure 8.4.

Let's also compute $P + P$, which we write as $2P$, as in any abelian group. (Note that this does not refer to the point $(0, 2)$, gotten by scalar multiplication of a vector, because $(0, 2)$ is not even on the curve E.) We start by computing $\lambda = 0$, so the equation for L is just $y = 1$. The third point of intersection has $x_3 = 0 - 0 - 0 = 0$, and then the y-coordinate must be -1. So $2P = (0, -1)$.

EXCERCISE: What is $3P$?

SOLUTION: By definition, $3P = 2P + P$, so that means we need to compute $(0, -1) + (0, 1)$. Our recipe says that this sum is $\mathcal{O}$, because $x_1 = x_2$ and $y_1 \neq y_2$.

Let's try working with the same equation, but now we'll use a finite field. We already computed Δ_E for $y^2 = x^3 + 1$, and saw that the only primes dividing Δ_E are 2 and 3. Therefore, $E(\mathbf{F}_7)$ is an abelian group.

EXCERCISE: List all of the elements of $E(\mathbf{F}_7)$. Don't forget $\mathcal{O}$.

SOLUTION: The simplest way to do this is to let x take on the value of each of the 7 elements in $\mathbf{F}_7$, and see if there is a y-value that solves the equation. For example, if $x = 0$, we get $y^2 = 0^3 + 1 = 1$, so $y = 1$ or $y = 6$ as an element of $\mathbf{F}_7$. We therefore obtain two points on $E(\mathbf{F}_7)$: $(0, 1)$ and $(0, 6)$. Continuing with $x = 1, 2, \ldots,$ we get $E(\mathbf{F}_7) = \{\mathcal{O}, (0, 1), (0, 6), (1, 3), (1, 4), (2, 3), (2, 4), (3, 0), (4, 3), (4, 4), (5, 0), (6, 0)\}$.

EXCERCISE: In $E(\mathbf{F}_7)$, what is $(0, 1) + (4, 4)$?

SOLUTION: We start by computing $\lambda = (4 - 1)/(4 - 0) = 3 \cdot 4^{-1} = 6$. The equation for L is therefore $y = 6x + 1$. We know that $x_3 = 6^2 - 0 - 4 = 4$, and then $P + Q = (4, 3)$.

EXCERCISE: Let's still continue using the equation $y^2 = x^3 + 1$. List all of the elements in $E(\mathbf{F}_5)$, and write out an addition table for the abelian group.

SOLUTION: The elements are $E(\mathbf{F}_5) = \{\mathcal{O}, (0, 1), (0, 4), (2, 2), (2, 3), (4, 0)\}$. The addition table is in table 8.1.

Table 8.1. Addition table for $E(\mathbf{F}_5)$

+	$\mathcal{O}$	(0,1)	(0,4)	(2,2)	(2,3)	(4,0)
$\mathcal{O}$	$\mathcal{O}$	(0,1)	(0,4)	(2,2)	(2,3)	(4,0)
(0,1)	(0,1)	(0,4)	$\mathcal{O}$	(2,3)	(4,0)	(2,2)
(0,4)	(0,4)	$\mathcal{O}$	(0,1)	(4,0)	(2,2)	(2,3)
(2,2)	(2,2)	(2,3)	(4,0)	(0,4)	$\mathcal{O}$	(0,1)
(2,3)	(2,3)	(4,0)	(2,2)	$\mathcal{O}$	(0,1)	(0,4)
(4,0)	(4,0)	(2,2)	(2,3)	(0,1)	(0,4)	$\mathcal{O}$

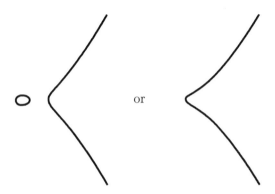

or

Figure 8.5. Two possible elliptic curves over $\mathbf{R}$

6. Topology

Although the content of this section is not needed elsewhere in this book, it is fun to see what kind of shapes are created by considering the real and complex points of an elliptic curve.

Let's start with a cubic polynomial $p(x)$ with three distinct roots of the form $p(x) = x^3 + Ax + B$. In the rest of this book, we suppose that A and B are integers, but in this section, they could be other kinds of real numbers. Let E be the projective curve defined by

$$y^2 z = x^3 + Axz^2 + Bz^3.$$

So E is an elliptic curve, and its identity element is the point $\mathcal{O}$ with projective coordinates $(0 : 1 : 0)$, which is the only point of E at infinity.

First let's see what shape is made by $E(\mathbf{R})$, that is, the solutions with real coordinates. We can see most of this by graphing $y^2 = x^3 + Ax + B$ in Cartesian coordinates. If we vary A and B we'll see that the picture always looks qualitatively like one of the two in figure 8.5. Either there is one oval and one curve that gets closer and closer to the vertical, or there is just the vertical-tending curve alone. Now remember that $\mathcal{O}$ is in $E(\mathbf{R})$ and lies at infinity on all vertical lines. When we throw in $\mathcal{O}$, we see that the vertical-tending curve goes through $\mathcal{O}$ both at the top and bottom in tinker-toy fashion. Compare the illustration on p. 55 of how the parabola fastens up into a loop.

In summary, we can say that $E(\mathbf{R})$ always consists of loops, either just one, or a pair of disjoint loops. Now why should this be so? Well, we know that E is a nonsingular curve, which means that at every point there is a single, well-defined tangent line to the curve. A tangent line to a curve is a good approximation to the curve, near the point of tangency, as we learn in calculus. So near any real point on the elliptic curve, $E(\mathbf{R})$ must look like a single small arc, a little bit curved of course. (A rigorous proof of this statement uses the implicit function theorem from calculus.)

Since $\mathcal{O}$ is a real point on the elliptic curve, there is at least one of these small arcs, and as we move along it, each new point assures us that it has an arc through it, which must continue the previous arc smoothly. So starting at any point of $E(\mathbf{R})$ we can move along the curve without ever coming to a point where we are forced to stop. The only way this can happen is if we go around in a loop, or we go around in a wiggly line without ever coming back to the same point. The latter possibility can be ruled out, using a topological property of the real projective plane called "compactness."

At this point, we know that $E(\mathbf{R})$ must consist of at least one loop, and if there is more than one loop, they must be disjoint. (This fact is true of $C(\mathbf{R})$ for any nonsingular real projective algebraic curve C.) But why are there always 1 or 2 loops, and not ever any other number? Let's use some more calculus to investigate this.

A loop in the plane must have two vertical tangents, one on each side of it. See figure 8.6.

The same is true of the "loop" that goes through the point at infinity on the end of the vertical lines. In this case, one tangent point is in the finite part of the plane, and the other is some vertical line. To find out which

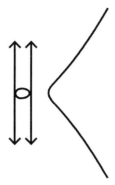

Figure 8.6. Two vertical tangents on a loop

one, you'd have to look in a different coordinate patch. It turns out to be the vertical line "all the way to the right," that is, the line at infinity.[2]

Now how many vertical tangents does $E(\mathbf{R})$ have? A vertical tangent will occur at $\mathcal{O}$ and at any point in the finite xy-plane where the partial derivative in the y-direction of $f(x, y) = y^2 - x^3 - Ax - B$ is 0. Compute: $f_y = 2y$. This is zero if and only if $y = 0$. So the vertical tangents occur at the points $(a, 0)$ where a is a solution of $x^3 + Ax + B = 0$. There are at most three real roots of the cubic. Together with $\mathcal{O}$ we get at most four vertical tangents. So there must be at most 2 loops.

Because nonreal roots of a polynomial with real coefficients must come in complex conjugate pairs, a real cubic polynomial always has 1 or 3 real roots (counted with multiplicities). If there are 3 real solutions of the equation $x^3 + Ax + B = 0$, we get two loops in $E(\mathbf{R})$, but if there is only 1 real solution, we only get 1 loop. In this case, there are two other "vertical" tangents, both tangent to nonreal points in $E(\mathbf{C})$.

Now let's move on to the topology of $E(\mathbf{C})$. This is more complex (pun intended) and we will have to be a bit sketchy. Reasoning similar to what we did for $E(\mathbf{R})$ will tell us that near any point in $E(\mathbf{C})$, the curve will look

---

[2] Let's do the computation. The homogeneous function defining E is $y^2z - x^3 - Axz^2 - Bz^3$. We want to find the tangent line to this curve at the point with projective coordinates $(0 : 1 : 0)$. So we need to look in the coordinate patch where $y \neq 0$. We can use homotheties to ensure that $y = 1$ in this coordinate patch. Thus the function defining E in this new xz-coordinate patch is $g(x, z) = z - x^3 - Axz^2 - Bz^3$. Work out its partial derivatives: $g_x = -3x^2 - Az^2$, $g_z = 1 - 2Axz - 3Bz^2$. At $(x, z) = (0, 0)$ these take the values $g_x = 0$ and $g_z = 1$. So in the xz-coordinate patch, the tangent line has the equation $0x + 1z = 0$, or $z = 0$. Since $z = 0$, this line lies completely at infinity in our original plane, which is the xy-coordinate patch. In other words, it is the line at infinity.

like a little bit of the complex plane, just as a little piece of $E(\mathbf{R})$ looks like a little bit of the real line. We can take each little bit of the complex plane to be an open disc. As we move from point to point along $E(\mathbf{C})$, these discs glue together smoothly and continuously and form a surface. This is called the *Riemann surface* determined by E.

Similarly to the case of $E(\mathbf{R})$, this Riemann surface can't come to an abrupt end or boundary anywhere. (It is "compact" and "without boundary.") Compare this to the Riemann surface determined by the complex projective line. When we glue up the finite line, which is the same thing as the complex numbers themselves, by adding a point at infinity, we get a sphere. (See p. 49.) What shape will we get for $E(\mathbf{C})$?

As far back as the early nineteenth century, mathematicians knew that $E(\mathbf{C})$ is topologically a torus, that is, the surface of a doughnut. It turns out that starting with an elliptic curve E, you can find a complex number τ and a function Z from $E(\mathbf{C})$ to $\mathbf{C}$ with the following properties:

- Let Π be the set of all points of the form $s + t\tau$ where s, t are real numbers satisfying the inequalities $0 \leq s, t \leq 1$. The parallelogram Π is called a *fundamental parallelogram* for E; see figure 8.7. The complex number $Z(x, y)$ must have a value in the fundamental parallelogram.
- We can make Z into a continuous one-to-one function by doing something strange with Π. We identify the left- and right-hand sides of the parallelogram, and we identify the top and bottom sides. In other words, if (x_0, y_0) and (x_1, y_1) are two nearby points on $E(\mathbf{C})$ (where nearby just means that x_0 is close to x_1 and y_0 is close to y_1), $Z(x_0, y_0)$ and $Z(x_1, y_1)$ are required to be close to each other in Π. In order to ensure that this happens, we need to say (by fiat) that points near to the bottom edge are close to points near the top edge, and similarly for the other two edges. This definition of "close" in turn forces a point near any particular vertex to be close to a point at any of the three other vertices.

Now, we can take Π and glue its edges together in parallel pairs. If we call the glued-up object $\widetilde{\Pi}$, then we can say that $E(\mathbf{C})$ is "isomorphic" to $\widetilde{\Pi}$. In fact, they are "complex analytically" isomorphic, and determine the same Riemann surface.

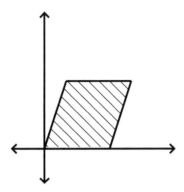

Figure 8.7. A fundamental parallelogram

Now get yourself a piece of flexible material, cut it into a parallelogram, and glue the parallel edges together. You will get a torus. We see $E(\mathbf{C})$ is topologically a torus. In fact it is called a "complex torus."

7. Other Important Facts about Elliptic Curves

The topology of elliptic curves is not particularly relevant for us as we attempt to explain the BSD Conjecture. We care much more about finite fields such as $\mathbf{F}_p$ and about the size of $E(\mathbf{F}_p)$ for a fixed curve E as p varies (always with the restriction that p does not divide Δ_E; see chapter 9 for what to do otherwise). If you looked at the exercises above, for the curve $y^2 = x^3 + 1$, you saw that $\#E(\mathbf{F}_5) = 6$ and $\#E(\mathbf{F}_7) = 12$. (For any finite set S, $\#S$ denotes the number of elements in S.) Moreover, if you actually worked the exercises yourself, you noticed that there were p possible values to substitute for x. In about half of the cases, there was no y-value, and in about half of the cases, there were 2 y-values. By chance, there happened to be a few x-values that produced only the single y-value of 0, but that's relatively rare, and insignificant when the modulus p gets large and there are lots of solutions.

Therefore, for a general cubic equation, given by equation (8.3), we expect there to be about p solutions to the congruence $y^2 \equiv x^3 + Ax + B$ (mod p). The elliptic curve contains the additional point $\mathcal{O}$, which is not a solution to that congruence, so we expect that the number of points in the group $E(\mathbf{F}_p)$ to be approximately $p + 1$. Let N_p be this number $\#E(\mathbf{F}_p)$.

Table 8.2. a_p for the curve $y^2 = x^3 + 1$

p	5	7	11	13	17	19	23	29	31	37
a_p	0	-4	0	2	0	8	0	0	-4	-10

It turns out that in fact $\#E(\mathbf{F}_p)$ must be within $2\sqrt{p}$ of $p + 1$, as stated in the following theorem.

THEOREM 8.5 (Hasse): The number N_p satisfies the inequalities $-2\sqrt{p} \leq p + 1 - N_p \leq 2\sqrt{p}$.

This inequality was conjectured by the German mathematician Emil Artin (1898–1962), and proved in 1933 by German mathematician Helmut Hasse (1898–1979). The quantity $p + 1 - N_p$ turns out to be a most important one for us, and we define

$$a_p = p + 1 - N_p.$$

WARNING: It is traditional to use a_p to stand for the quantity $p + 1 - N_p$, even though we also used the symbols a_1, a_2, a_3, a_4, and a_6 above to stand for constants in equation (8.2). It is very rare that there will be any ambiguity about which numbers are meant.

Hasse's Theorem says that $|a_p| \leq 2\sqrt{p}$. Notice that a_p is only defined here for those primes p for which p does not divide Δ_E. In chapter 9, we'll talk about how to define a_p in the remaining cases.

These numbers a_p measure how far N_p deviates from the intuitive guess of $p + 1$. There are computer programs that can be used to calculate them rapidly. In table 8.2, we list some of them for the curve $y^2 = x^3 + 1$.

Finally, one more theorem. If p is any prime, and n any positive integer, recall that there is a field with p^n elements, which is written $\mathbf{F}_{p^n}$. (In section 7 of chapter 2, we saw one way to construct the field $\mathbf{F}_4$ and mentioned the general procedure.) If p does not divide Δ_E, then E is also nonsingular when we allow the variables to take values in $\mathbf{F}_{p^n}$ for any n. Let N_{p^n} be the number of elements in $E(\mathbf{F}_{p^n})$. The following theorem tells us that the numbers N_{p^n} obey inequalities similar to those in theorem 8.5.

Table 8.3. $x^3 - x$ in $\mathbf{F}_9$

x	0	1	2	i	$2i$	$1+i$	$1+2i$	$2+i$	$2+2i$
x^3	0	1	2	$2i$	i	$1+2i$	$1+i$	$2+2i$	$2+i$
$x^3 - x$	0	0	0	i	$2i$	i	$2i$	i	$2i$

THEOREM 8.6: $-2\sqrt{p^n} \le p^n + 1 - N_{p^n} \le 2\sqrt{p^n}$. Moreover, if $n > 1$, then there is a formula to compute N_{p^n} in terms of N_p.

WARNING: We will also discuss numbers a_{p^n} later on. In general, a_{p^n} is *not* the same as $p^n + 1 - N_{p^n}$.

8. Two Numerical Examples

Theorem 8.6 is sufficiently interesting that we'll work a couple of numerical examples now. Later on, we'll explain how to compute N_{p^n} using N_p without actually having to count points in $E(\mathbf{F}_{p^n})$.

For our first example, let's let E be defined by the equation $y^2 = x^3 - x$. You can compute $\Delta_E = 64$, so it is not a problem to discuss the group $E(\mathbf{F}_3)$. A very little bit of computation shows that $E(\mathbf{F}_3) = \{\mathcal{O}, (0,0), (1,0), (2,0)\}$, where all of the points other than $\mathcal{O}$ have order 2. We see that $N_3 = 4$, and $a_3 = 0$.

In order to talk about $E(\mathbf{F}_9)$, we need to discuss the field $\mathbf{F}_9$ briefly. There are many ways to describe it, and the simplest is

$$\mathbf{F}_9 = \left\{ a + bi \mid a, b \in \mathbf{F}_3,\ i^2 = 2 \right\}.$$

WARNING: We are using the letter i here because $i^2 = 2 = -1$ in $\mathbf{F}_9$. Until we finish this example, this letter i is an element of $\mathbf{F}_9$ and not an element of $\mathbf{C}$.

You can check that in $\mathbf{F}_9$, $(a + bi)^3 = a - bi$. You can also check that the squares in $\mathbf{F}_9$ are 0, 1, $2 = i^2 = (2i)^2$, $2i = (1+i)^2 = (2+2i)^2$, and $i = (2+i)^2 = (1+2i)^2$. We can make a little chart of $x^3 - x$ for the 9 elements of $\mathbf{F}_9$ in table 8.3.

Table 8.4. $t^3 - t$ and $t^2 - t$ in $\mathbf{F}_4$

t	0	1	α	$1 + \alpha$
$t^3 - t$	0	0	$\alpha + 1$	α
$t^2 - t$	0	0	1	1

Every x-value therefore has at least one y-value that makes $x^3 - x = y^2$, and each x-value other than 0, 1, and 2 has 2 different y-values. Therefore, $\#E(\mathbf{F}_9) = 16$ (don't forget to include $\mathcal{O}$ when counting points). You can check that $p^n + 1 - N_{p^n} = 3^2 + 1 - 16 = -6$, while the lower bound in Theorem 8.6 is $-2\sqrt{9}$, which is also -6. We'll mention that $a_9 = -3$, and we'll explain later how to define and compute a_9.

Let's work one more example. Because we wrote out the addition and multiplication tables for $\mathbf{F}_4$ in table 2.2 on page 40, we'll work in $\mathbf{F}_2$ and $\mathbf{F}_4$. In order to construct an elliptic curve E for which Δ_E is not a multiple of 2, we need to use the more complicated formula (8.2) to define E. If we use the equation $y^2 - y = x^3 - x$, some computations tell us that $\Delta_E = 37$. It is therefore acceptable to discuss $E(\mathbf{F}_2)$ (or, for that matter, $E(\mathbf{F}_p)$ as long as $p \neq 37$).

A bit of work tells us that $E(\mathbf{F}_2) = \{\mathcal{O}, (0, 0), (0, 1), (1, 0), (1, 1)\}$. Therefore, $N_2 = 5$, and $a_2 = 2 + 1 - 5 = -2$. The lower bound in theorem 8.5 is $-2\sqrt{2} \approx -2.8$, so we verify again that the bound works.

Remember that in $\mathbf{F}_4$ (as well as any other field which has characteristic 2), addition and subtraction are the same thing. In $\mathbf{F}_4$, we can compute using table 2.2 that $x^3 = 1$ if $x \neq 0$, and of course $0^3 = 0$. We compute in table 8.4 the values of $t^3 - t$ and $t^2 - t$ in $\mathbf{F}_4$.

In this case, we see that $E(\mathbf{F}_4) = E(\mathbf{F}_2)$, and then $4 + 1 - N_4 = 5 - 5 = 0$. This computation certainly agrees with the inequalities in Theorem 8.6. We mention that $a_4 = 2$ in this case, and we'll explain later how to verify this equation.

Chapter 9

·　·　◈　·　·

SINGULAR CUBICS

Road Map

Let E be an elliptic curve defined by a cubic equation with integer coefficients. The details of the BSD Conjecture require us also to work with singular cubic equations, because there will always be some prime numbers p dividing the discriminant Δ_E. If we reduce the equation for E modulo such p, we will get a singular cubic, and we will need to count the number of points on these singular curves as well.

Our goal is to show that the group law on a singular cubic curve is always "the same as"—in technical terms, "isomorphic to"—a group law that can be defined in some other, easier-to-understand way. The reasoning is intricate, and a lot of the messy algebraic details can be skipped without loss as long as you are willing to believe the conclusion. In the course of this chapter, we produce various rational parametrizations of the nonsingular points of a singular cubic curve. It may be helpful if you compare this with the parametrization of the circle we derived in the Prologue.

1. The Singular Point and the Group Law

Let $\overline{C}$ be a projective cubic curve defined by the homogeneous equation $G(X, Y, Z) = 0$. In this section, we will be working with the part of $\overline{C}$ which lies in the finite plane, so in this chapter we return to using the notation $\overline{C}$ to stand for the projective curve, and C to represent the part of $\overline{C}$, that lies in the finite plane. We will suppose that the coefficients of the equation and the allowable values of X, Y, and Z are elements of a field K. We will further suppose that the characteristic of K is neither 2 nor 3 throughout this chapter. In other words, we stipulate that neither $\mathbf{F}_2$ nor $\mathbf{F}_3$ are subfields of K. This restriction allows us to divide by 2 or 3 in the arguments that follow. There is an analysis of the situation for fields of characteristic 2 or 3, but the arguments are even more complicated than the ones which we will now present.

If $\overline{C}$ is a singular curve, then there must by definition be a singular point P. You can see an illustration of the real points of two singular cubics, along with the point P, in figure 4.4. Any line L which passes through the point P must pass through with multiplicity at least 2. Suppose that Q is another singular point, and let L be the line connecting P and Q. Then the line L passes through each of those two points P and Q with multiplicity at least 2, which implies that L intersects $\overline{C}$ with multiplicity at least 4. This is not possible, because $\overline{C}$ is a cubic and so any line can intersect $\overline{C}$ in at most 3 points. Our conclusion is that *a singular cubic curve $\overline{C}$ contains only one point of singularity*, which we will call P. We will use the notation C_{ns} to refer to all points of $\overline{C}$ except for P (the subscript "ns" is short for "nonsingular"). Steps 1 and 2 in section 1 of chapter 8 still work to define a group law on C_{ns}. The main thing is that if Q_1 and Q_2 are nonsingular points on C, then the line connecting Q_1 and Q_2 cannot go through P. If it did, it would have global intersection multiplicity at least 4, and C has degree 3.

2. The Coordinates of the Singular Point

We will choose a coordinate patch containing P, and write the equation defining $\overline{C}$ on this patch in the form $F(X, Y) = 0$. Suppose that the coordinates of the point P are (α, β). The definition of singular point (p. 80 or p. 96) implies that we must have

$$F_X(\alpha, \beta) = F_Y(\alpha, \beta) = 0.$$

We may suppose that we can write the equation for C (the part of $\overline{C}$ contained in this patch) in the form

$$X^3 + A_2 X^2 + A_4 X + A_6 - Y^2 = 0 \tag{9.1}$$

with A_2, A_4, and A_6 in K. This special form can always be arranged for any cubic, singular or not, provided that $1 + 1 \neq 0$ in the field K. The result of some algebraic transformations (see an appendix to this chapter for the gory details) is that equation (9.1) becomes

$$y^2 = x^3 + a_2 x^2 + a_4 x + a_6,$$

where $a_2, a_4, a_6 \in K$, *and* $(0,0)$ is the singular point P. Therefore, the equation $x^3 + a_2 x^2 + a_4 x + a_6 = 0$ has 0 as a root of multiplicity at least 2, which implies that $a_4 = a_6 = 0$. We rewrite the formula for C in the form

$$y^2 = x^3 + Ax^2. \tag{9.2}$$

Our analysis will now depend critically on the value of A.

We call the different cases "reductions" because they occur especially when we reduce the equation of an elliptic curve modulo a prime. In "additive" reduction, the group law on C_{ns} is isomorphic to the additive group $(K, +)$, and similarly for "multiplicative" reduction.

3. Additive Reduction

Suppose that $A = 0$. Equation (9.2) becomes $y^2 = x^3$. The real points of this equation are pictured in figure 4.6. We can define a function $\psi : K \to C_{\text{ns}}(K)$ with the formulas

$$\psi(t) = \begin{cases} (t^{-2}, t^{-3}) & \text{if } t \neq 0 \\ \mathcal{O} & \text{if } t = 0 \end{cases} \tag{9.3}$$

where $\mathcal{O}$ is the point at infinity. It is easy to check that the inverse function $t = \psi^{-1}(x, y)$ is defined by

$$\psi^{-1}(x, y) = \frac{x}{y}$$

$$\psi^{-1}(\mathcal{O}) = 0.$$

Table 9.1. Computing $\psi(t)$ in $\mathbf{F}_7$ for additive reduction

t	0	1	2	3	4	5	6
$\psi(t)$	$\mathcal{O}$	$(1,1)$	$(2,1)$	$(4,6)$	$(4,1)$	$(2,6)$	$(1,6)$

The reason to introduce this function is that ψ is not just a one-to-one correspondence between K and $C_{ns}(K)$; it is an *isomorphism*. In general, an isomorphism between two objects is a function that shows that the "essential structure" of the two objects is the same. In this case, ψ is an isomorphism of groups.

Our claim that ψ is an isomorphism means first that ψ is a one-to-one function with an inverse. That is easy to check, because we showed you a formula for ψ^{-1}. The second part of the claim is harder to check. It says that if t and u are any elements of K, then

$$\psi(t + u) = \psi(t) + \psi(u). \tag{9.4}$$

Here the addition on the right-hand side of equation (9.4) means addition of points on the singular cubic curve. Equation (9.4) is therefore saying something very interesting.

We hide the details of the proof of equation (9.4) in an appendix at the end of this chapter, and instead we give here an example of what it says. Let's take the particular case where $K = \mathbf{F}_7$, the field of 7 elements. The function ψ defined in (9.3) takes these values given in table 9.1. Because $4 + 5 = 2$ in $\mathbf{F}_7$, the isomorphism says that $(4, 1) + (2, 6) = (2, 1)$ added as points on C_{ns}.

EXERCISE: Verify that indeed $(4, 1) + (2, 6) = (2, 1)$ on C_{ns}.

SOLUTION: The line connecting $(4, 1)$ and $(2, 6)$ is $y = x + 4$. Substitute this equation into $y^2 = x^3$, and we get $(x + 4)^2 = x^3$. Expanding yields $x^3 - x^2 - x - 2 = 0$. Because the coefficient of x^2 is -1, the 3 roots of this equation add up to 1, so the third root must be $x = 2$. Finally, we compute the y-value on this line corresponding to $x = 2$, giving $y = 6$, and negate it, giving $(2, 1)$, as desired.

4. Split Multiplicative Reduction

Return to equation (9.2). Suppose next that $A = B^2 \neq 0$ for some $B \in K$. We will need to make another simplification, so rewrite the equation temporarily using uppercase letters:

$$Y^2 = X^3 + B^2 X^2.$$

Divide by B^6, and the equation becomes

$$\left(\frac{Y}{B^3}\right)^2 = \left(\frac{X}{B^2}\right)^3 + \left(\frac{X}{B^2}\right)^2.$$

Now we let $y = Y/B^3$ and $x = X/B^2$, and we arrive at

$$y^2 = x^3 + x^2. \tag{9.5}$$

The real points of this equation are pictured in figure 4.6. This case is called "split," because the two tangential lines at the singular point P split apart over the field K; we do not have to use a larger field to split them.

Define a function $\phi : C_{ns}(K) \to K^\times$ with the formulas

$$\phi(x, y) = \frac{y + x}{y - x}$$

$$\phi(\mathcal{O}) = 1.$$

(Recall that $K^\times$ is the notation for $K - \{0\}$, the set of nonzero elements in the field K, with group operation multiplication.)

We first need to verify that the function ϕ is defined properly. To start, we need to make sure that we are not dividing by 0 in the formula. If $y - x = 0$, then $x = y$, and so $y^2 = x^2$. Combined with the formula $y^2 = x^3 + x^2$, we can deduce that $x^3 = 0$, and then $(x, y) = (0, 0)$, which is the singular point on C. Because our function is defined for $C_{ns}(K)$ and not C, we have no problem.

We also need to make sure that the function ϕ takes only nonzero values. If $\phi(x, y) = 0$, then $y + x = 0$, which implies that $y = -x$, and then $y^2 = x^2$. We can conclude again that $(x, y) = (0, 0)$.

We next want to show that the function ϕ is an isomorphism. To show that it is a one-to-one correspondence between the points of $C_{ns}(K)$ and the elements of $K^\times$, we need to find the inverse of ϕ. Let $t = \frac{y+x}{y-x}$, and let's solve for x in terms of t:

$$t = \frac{y+x}{y-x}$$

$$t(y-x) = y+x$$

$$ty - y = x + tx$$

$$y(t-1) = x(t+1)$$

$$y^2(t-1)^2 = x^2(t+1)^2$$

$$(x^3 + x^2)(t-1)^2 = x^2(t+1)^2$$

$$(x+1)(t-1)^2 = (t+1)^2$$

$$x + 1 = \frac{(t+1)^2}{(t-1)^2}$$

$$x = \frac{(t+1)^2}{(t-1)^2} - 1 = \frac{(t+1)^2 - (t-1)^2}{(t-1)^2} = \frac{4t}{(t-1)^2}$$

$$y = x\frac{t+1}{t-1} = \frac{4t(t+1)}{(t-1)^3}$$

So our formula for ϕ^{-1} becomes

$$\phi^{-1}(t) = \begin{cases} \left(\dfrac{4t}{(t-1)^2}, \dfrac{4t(t+1)}{(t-1)^3} \right) & t \neq 1 \\ \mathcal{O} & t = 1. \end{cases} \tag{9.6}$$

We have now shown that ϕ gives a one-to-one correspondence. The more interesting (and harder!) part is showing that the group laws interact properly with the function ϕ. Rather than work with the function ϕ, it is simpler to let $\psi = \phi^{-1}$, and use formulas involving ψ.

We must show that

$$\psi(tu) = \psi(t) + \psi(u). \tag{9.7}$$

Table 9.2. Computing $\psi(t)$ for split multiplicative reduction

t	1	2	3	4	5	6
$\psi(t)$	$\mathcal{O}$	(1,3)	(3,6)	(1,4)	(3,1)	(6,0)

Equation (9.7) is slightly different from equation (9.4) because the group operation in the group $K^\times$ is multiplication rather than addition.

We again put the details of the proof of (9.7) in an appendix to this chapter, and work through an example instead. The defining equation of our curve is $y^2 = x^3 + x^2$, and again we work in the field $\mathbf{F}_7$. Now the function ψ in table 9.2. Because $4 \cdot 5 = 6$ in $\mathbf{F}_7^\times$, the isomorphism tells us that $(1, 4) + (3, 1) = (6, 0)$ on C_{ns}.

EXERCISE: Verify that $(1, 4) + (3, 1) = (6, 0)$ on C_{ns}.

SOLUTION: The equation of the line through the points $(1, 4)$ and $(3, 1)$ is $y = 2x + 2$. Substitution into $y^2 = x^3 + x^2$ yields $x^3 - 3x^2 - x - 4 = 0$. Looking at the coefficient of x^2, we see that the 3 roots of the equation sum to 3, meaning that the third root must be $x = 6$. Substitution of $x = 6$ into the equation of the line yields $y = 0$, which when negated is again 0, so the sum is indeed $(6, 0)$.

5. Nonsplit Multiplicative Reduction

In order to describe the isomorphism in the remaining case, where A is not the square of an element of K, we need to tell you about another group. If A is not a square in K, we define the field $K(\sqrt{A})$ with the equation

$$K(\sqrt{A}) = \{x + y\sqrt{A} \ : \ x, y \in K\}.$$

This case is called "nonsplit" because the two tangential lines at the singular point do not split over K, but only over the field $K(\sqrt{A})$. An example of this over $\mathbf{R}$ is $y^2 = x^3 - x^2$, which is pictured in figure 9.1. Note that $(0, 0)$ is an isolated point on this graph, and therefore it is impossible to define any real tangential lines at that point. Some of the algebra in this section will be reminiscent of the computations in the appendix to chapter 7.

Figure 9.1. $y^2 = x^3 - x^2$

DEFINITION: If $t \in K(\sqrt{A})$ and $t = x + y\sqrt{A}$, we define $\bar{t} = x - y\sqrt{A}$. We call $\bar{t}$ the *Galois conjugate* of t. The number $t\bar{t}$, which is $x^2 - Ay^2$, is called the *norm* of t, written $\mathcal{N}(t)$. We define the set U with the equation

$$U = \{t \in K(\sqrt{A}) \,:\, t\bar{t} = 1\}.$$

The symbol U is short for the word "unit."

The field $\mathbf{C} = \mathbf{R}(\sqrt{-1})$ is a familiar example of these ideas. The word "conjugate" is meant to remind you of complex conjugation. Just as with complex conjugation, it is true that $\overline{tu} = \bar{t}\bar{u}$ and $\overline{t + u} = \bar{t} + \bar{u}$. Notice that $\mathcal{N}(t) \in K$, and $\mathcal{N}(t) = 0$ if and only if $t = 0$. Also notice that

$$t \in K \text{ if and only if } t = \bar{t}. \tag{9.8}$$

The set U is closed under multiplication; in fact, U is an abelian group.[1] Moreover,

$$t \in U \text{ if and only if } \bar{t} = t^{-1}. \tag{9.9}$$

If C is defined by (9.2), where A is not a square, we define $\phi : C_{\mathrm{ns}}(K) \to U$ with the formulas

$$\phi(x, y) = \frac{y + x\sqrt{A}}{y - x\sqrt{A}}$$

$$\phi(\mathcal{O}) = 1.$$

We first need to check that this definition makes sense, and then we can explain that it defines an isomorphism.

---

[1] Compare pp. 112–115.

To begin, we need to make sure that $\phi(x, y)$ is defined. The problem we need to avoid is division by 0. If $y - x\sqrt{A} = 0$, then $y = x\sqrt{A}$, and then $y^2 = Ax^2$. We also have $y^2 = x^3 + Ax^2$. These two equations together imply $x = 0$, which in turn implies $y = 0$. Because $(0, 0)$ is not in C_{ns}, we know that $y - x\sqrt{A} \neq 0$.

Next, we need to make sure that $\phi(x, y) \in U$. Using properties of the norm, we have

$$\mathcal{N}\left(\frac{y + x\sqrt{A}}{y - x\sqrt{A}}\right) = \frac{\mathcal{N}(y + x\sqrt{A})}{\mathcal{N}(y - x\sqrt{A})} = \frac{y^2 - Ax^2}{y^2 - Ax^2} = 1.$$

Next, we can show that the function ϕ is one-to-one and onto by finding its inverse. Write $t = \phi(x, y)$. We then have

$$t = \frac{y + x\sqrt{A}}{y - x\sqrt{A}}$$

$$t(y - x\sqrt{A}) = y + x\sqrt{A}$$

$$ty - y = x\sqrt{A} + tx\sqrt{A}$$

$$y(t - 1) = x\sqrt{A}(t + 1)$$

$$y^2(t - 1)^2 = Ax^2(t + 1)^2$$

$$(x^3 + Ax^2)(t - 1)^2 = Ax^2(t + 1)^2$$

$$(x + A)(t - 1)^2 = A(t + 1)^2$$

$$x + A = A\frac{(t + 1)^2}{(t - 1)^2}$$

$$x = A\left(\frac{(t + 1)^2}{(t - 1)^2} - 1\right)$$

$$= A\left(\frac{(t + 1)^2 - (t - 1)^2}{(t - 1)^2}\right) = \frac{4At}{(t - 1)^2} \qquad (9.10)$$

$$y = x\sqrt{A}\frac{t + 1}{t - 1} = \frac{4A\sqrt{A}t(t + 1)}{(t - 1)^3} \qquad (9.11)$$

Table 9.3. $y^2 = x^3 + 3x^2$ over $\mathbf{F}_7$

t	1	6	$2+\sqrt{3}$	$2+6\sqrt{3}$	$3\sqrt{3}$	$4\sqrt{3}$	$5+\sqrt{3}$	$5+6\sqrt{3}$
$\psi(t)$	$\mathcal{O}$	(4,0)	(6,4)	(6,3)	(1,5)	(1,2)	(5,2)	(5,5)

We are not yet done. We showed above that if $(x, y) \in C_{ns}(K)$, then $\phi(x, y) = t \in U$. Now, we need to show that if $t \in U$, then x and y as defined in (9.10) and (9.11) are elements of K. We will do this by using (9.9) to show that $x = \bar{x}$ and $y = \bar{y}$ and then relying on (9.8).

We have

$$\bar{x} = \frac{\overline{4At}}{\overline{(t-1)^2}} = \frac{4A\bar{t}}{(\bar{t}-1)^2} = \frac{4At^{-1}}{(t^{-1}-1)^2}$$

$$= \left(\frac{4At^{-1}}{(t^{-1}-1)^2}\right)\left(\frac{t^2}{t^2}\right) = \frac{4At}{(1-t)^2} = x$$

$$\bar{y} = \overline{x\sqrt{A}}\frac{\overline{t+1}}{t-1} = -x\sqrt{A}\frac{\bar{t}+1}{\bar{t}-1} = -x\sqrt{A}\frac{t^{-1}+1}{t^{-1}-1}$$

$$= -x\sqrt{A}\frac{1+t}{1-t} = x\sqrt{A}\frac{t+1}{t-1} = y$$

Similar to (9.6), we have

$$\phi^{-1}(t) = \begin{cases} \left(\dfrac{4At}{(t-1)^2}, \dfrac{4A\sqrt{A}t(t+1)}{(t-1)^3}\right) & t \neq 1 \\ \mathcal{O} & t = 1 \end{cases}$$

Again, we let $\psi = \phi^{-1}$. We again need to show that $\psi(tu) = \psi(t) + \psi(u)$, which is identical to (9.7), and again the details go into an appendix.

Our example curve this time is $y^2 = x^3 + 3x^2$ with the field $\mathbf{F}_7$. Notice that 3 is not a square in $\mathbf{F}_7$, so this is indeed an example of nonsplit multiplicative reduction. The function ψ is a bit trickier to understand, because it has U as its domain. You can compute that we get table 9.3. For example, we can compute that $(5 + 6\sqrt{3})(4\sqrt{3}) = 2 + 6\sqrt{3}$ (remembering as always that we take our coefficients in $\mathbf{F}_7$), and therefore $(5, 5) + (1, 2) = (6, 3)$ on $C_{ns}(\mathbf{F}_7)$.

EXERCISE: Verify that $(5, 5) + (1, 2) = (6, 3)$ on $y^2 = x^3 + 3x^2$.

SOLUTION: The equation of the line through $(5, 5)$ and $(1, 2)$ is $y = 6x + 3$. Substitution into $y^2 = x^3 + 3x^2$ yields $x^3 - 5x^2 - x - 2 = 0$. The 3 roots of this equation sum to 5, and therefore the third root is $x = 6$. Substitution of $x = 6$ into $y = 6x + 3$ yields $y = 4$, and therefore the third point is $(6, 3)$.

We've done a lot of algebra in this section. It shows that the details of the group structure of a singular cubic curve can be fairly complicated. The details of an elliptic curve (a nonsingular cubic) are even more complicated. We should be gaining some respect for the powers of abstraction and for the great mathematicians who have made the theory more manipulable and conceptual than it would be if approached with brute force.

6. Counting Points

We pause to collect some information that is easily available from our computations above, and which is one of the key reasons for having performed them.

We saw above that on the curve $y^2 = x^3$, the group $C_{ns}(K)$ is isomorphic to the additive group of K. In particular, if K is $\mathbf{F}_p$, the field with p elements, then $C_{ns}(\mathbf{F}_p)$ is isomorphic to the additive group of $\mathbf{F}_p$, and so the number of elements in $C_{ns}(\mathbf{F}_p)$ is p.

On the curve $y^2 = x^3 + x^2$, the group $C_{ns}(K)$ is isomorphic to $K^\times$, the multiplicative group of K. In particular, if K is $\mathbf{F}_p$, the field with p elements, then $C_{ns}(\mathbf{F}_p)$ is isomorphic to the multiplicative group $\mathbf{F}_p^\times$, and so the number of elements in $C_{ns}(\mathbf{F}_p)$ is $p - 1$.

On the curve $y^2 = x^3 + Ax^2$, where A is not a square in K, the group $C_{ns}(K)$ is isomorphic to U, the unit group of $K(\sqrt{A})$, which we defined in section 5. In particular, if K is $\mathbf{F}_p$, the field with p elements, then $C_{ns}(\mathbf{F}_p)$ is isomorphic to a group that contains $p + 1$ elements. (This is not quite trivial to prove, but at least you can verify that in the example we used before, where $p = 7$, the group has 8 elements.)

We will be most concerned with the number $a_p = p - N_p$. Why is the formula for a_p different for singular curves than for nonsingular

Table 9.4. Counting points on singular cubics

Type of reduction	N_p	a_p
Additive	p	0
Split Multiplicative	$p-1$	1
Nonsplit Multiplicative	$p+1$	-1

ones? Rather than $p + 1 - N_p$, we compute $p - N_p$, because we eliminate the singular point from our count. We summarize our computations in table 9.4.

7. Conclusion

Although the algebra in the preceding few pages has been quite complex, the essential point should not be missed. Indeed, in teaching this subject, one could assign each of these computations as an exercise, summarizing only the conclusion in class.

The important fact about all of these computations is that the group law on the points $C_{ns}(K)$ of a singular cubic curve C is the same as the group law of another, better-known group. The correspondence might be extremely messy and hard to compute. In the end, however, we can conclude that $C_{ns}(K)$ is the same as either an additive group K, a multiplicative group $K^\times$, or else a slightly more complicated multiplicative group U. In all cases, these are previously known objects, that is, ones that were understood prior to the study of elliptic curves. In contrast, the group law on an elliptic curve does not reduce to any other known object; it has its own structure and complexity. In particular, the number N_p of elements on an elliptic curve over $\mathbf{F}_p$ is *much* subtler than it is for a singular cubic curve. One reason for this is that a singular cubic has its singular point as a "handle" we can grab to begin our analysis, whereas a nonsingular cubic has no such "handle."

Appendix A: Changing the Coordinates of the Singular Point

We have seen that if P is the singular point, with coordinates (α, β), then we have $F_X(\alpha, \beta) = F_Y(\alpha, \beta) = 0$. The equation $F_Y(\alpha, \beta) = 0$ tells us that $-2\beta = 0$, which of course means that $\beta = 0$.

Write $f(X) = X^3 + A_2X^2 + A_4X + A_6$. The equation $F_X(\alpha, \beta) = 0$ combines with the equation $\beta = 0$ to tell us that $f'(\alpha) = 0$. The fact that

$F(\alpha, \beta) = 0$ means that $f(\alpha) = 0$. The two equations $f(\alpha) = f'(\alpha) = 0$ mean that α is a *root of multiplicity at least 2* of the polynomial $f(X)$, or that $f(X)$ is divisible by $(X - \alpha)^2$.

We wish to show that $\alpha \in K$, and the argument to do this is involved but elementary. Assume that $\alpha \notin K$. Divide $f(X)$ by $f'(X)$ using polynomial division. We get $f(X) = q(X)f'(X) + r(X)$, where the polynomials $q(X)$ and $r(X)$ both have coefficients in K, and the degree of $r(X)$ is 0 or 1 (because the degree of $f'(X)$ is 2).

Substitute $X = \alpha$, and we get $r(\alpha) = 0$. We know that $r(X)$ cannot be a nonzero multiple of $X - \alpha$, because $\alpha \notin K$. We also know that $r(X)$ is either a linear polynomial or a constant. We can conclude that $r(X)$ is the 0 polynomial. Therefore, $f'(X)$ divides $f(X)$.

We can write $f(X) = (X - \alpha)^2(X - \gamma)$, and $f'(X) = 3(X - \alpha)(X - \delta)$. Because $f'(X)$ divides $f(X)$, we can conclude that $\gamma = \delta$ or $\gamma = \alpha$. Now we can differentiate $f(X) = (X - \alpha)^2(X - \gamma)$ and directly see that the derivative is not $3(X - \alpha)(X - \gamma)$ unless $\alpha = \gamma$.

We can therefore deduce that $f(X) = (X - \alpha)^3$. The coefficient of the X^2-term in $f(X)$ is -3α, and which means that $-3\alpha \in K$. Because we are assuming that 3 is invertible in K, we can conclude that $\alpha \in K$.

Now we can let $X = x + \alpha$ and let $Y = y$, and then the singular point P has coordinates $x = 0$ and $y = 0$.

Appendix B: Additive Reduction in Detail

Verifying (9.4) is a matter of checking various cases. We will just check the hardest case, which is the general one. In that case, t and u are unequal and t, u, and $t + u$ are all nonzero, and there really *is* something magical going on. The left-hand side of the equation becomes $\psi(t + u) = ((t + u)^{-2}, (t + u)^{-3})$. The right-hand side becomes $(t^{-2}, t^{-3}) + (u^{-2}, u^{-3})$. Suppose that the sum of those two points is (v^{-2}, v^{-3}). We must show that $v = u + t$.

The slope of the line connecting the points (t^{-2}, t^{-3}) and (u^{-2}, u^{-3}) is $\frac{u^{-3} - t^{-3}}{u^{-2} - t^{-2}}$. Call this number λ. The equation of the line between (t^{-2}, t^{-3}) and (u^{-2}, u^{-3}) is $y - t^{-3} = \lambda(x - t^{-2})$, or $y = \lambda(x - t^{-2}) + t^{-3}$. Substitute into $x^3 - y^2 = 0$, and we have

$$x^3 - (\lambda(x - t^{-2}) + t^{-3})^2 = 0. \tag{9.12}$$

The coefficient of the x^2-term in equation (9.12) is $-\lambda^2$. The three roots of (9.12) are t^{-2}, u^{-2}, and v^{-2}. We therefore have $t^{-2} + u^{-2} + v^{-2} = \lambda^2$, or $v^{-2} = \lambda^2 - u^{-2} - t^{-2}$. Now we work out the algebra:

$$v^{-2} = \lambda^2 - u^{-2} - t^{-2} = \frac{(u^{-3} - t^{-3})^2}{(u^{-2} - t^{-2})^2} - (u^{-2} + t^{-2})$$

$$= \frac{(u^{-3} - t^{-3})^2}{(u^{-2} - t^{-2})^2} - \frac{(u^{-2} + t^{-2})(u^{-2} - t^{-2})^2}{(u^{-2} - t^{-2})^2}$$

$$= \frac{u^{-6} - 2u^{-3}t^{-3} + t^{-6}}{(u^{-2} - t^{-2})^2} - \frac{u^{-6} - t^{-2}u^{-4} - t^{-4}u^{-2} + t^{-6}}{(u^{-2} - t^{-2})^2}$$

$$= \frac{t^{-2}u^{-4} - 2u^{-3}t^{-3} + t^{-4}u^{-2}}{(u^{-2} - t^{-2})^2} = t^{-2}u^{-2}\frac{u^{-2} - 2u^{-1}t^{-1} + t^{-2}}{(u^{-2} - t^{-2})^2}$$

$$= t^{-2}u^{-2}\frac{(u^{-1} - t^{-1})^2}{(u^{-2} - t^{-2})^2} = \left(\frac{u^{-1}t^{-1}}{u^{-1} + t^{-1}}\right)^2 = \left(\frac{1}{u + t}\right)^2$$

implying that

$$v^2 = (u + t)^2. \tag{9.13}$$

We hope to conclude that $v = u + t$, but equation (9.13) allows for the possibility that $v = -(u + t)$. In order to finish the argument, we need to compute v^{-3}, which is the negative of the y-coordinate of the third point of intersection:

$$y = \lambda(x - t^{-2}) + t^{-3}$$

$$= \lambda(v^{-2} - t^{-2}) + t^{-3}$$

$$= \frac{u^{-3} - t^{-3}}{u^{-2} - t^{-2}}\left(\frac{1}{(u + t)^2} - \frac{1}{t^2}\right) + \frac{1}{t^3}$$

$$= \frac{t^3 - u^3}{(ut)(t^2 - u^2)}\left(\frac{t^2 - (u + t)^2}{t^2(u + t)^2}\right) + \frac{1}{t^3} \tag{9.14}$$

$$= -\frac{1}{(u + t)^3} \tag{9.15}$$

This tells us that $v^3 = (u + t)^3$, and finally that allows us to conclude that $v = u + t$, as claimed. We admit that we used a computer-algebra system at the end to go from (9.14) to (9.15). It is common and accepted these days to use such tools.

Appendix C: Split Multiplicative Reduction in Detail

The most interesting case of (9.7), and the only one we will work out, is when neither t nor u are 1; t and u are unequal; and $tu \neq 1$. We compute that

$$(x_1, y_1) = \psi(t) = \left(\frac{4t}{(t-1)^2}, \frac{4t(t+1)}{(t-1)^3} \right)$$

$$(x_2, y_2) = \psi(u) = \left(\frac{4u}{(u-1)^2}, \frac{4u(u+1)}{(u-1)^3} \right)$$

The line through (x_1, y_1) and (x_2, y_2) has slope

$$\lambda = \frac{y_2 - y_1}{x_2 - x_1} = \frac{\frac{4u(u+1)}{(u-1)^3} - \frac{4t(t+1)}{(t-1)^3}}{\frac{4u}{(u-1)^2} - \frac{4t}{(t-1)^2}}$$

$$= \frac{(4u)(u+1)(t-1)^3 - 4t(t+1)(u-1)^3}{(4u)(u-1)(t-1)^3 - 4t(t-1)(u-1)^3}. \tag{9.16}$$

The equation of the line through (x_1, y_1) and (x_2, y_2) is $y = \lambda(x - x_1) + y_1$. Substitute into (9.5), and we have $(\lambda(x - x_1) + y_1)^2 = x^3 + x^2$, or

$$0 = x^3 + x^2 - (\lambda(x - x_1) + y_1)^2. \tag{9.17}$$

In equation (9.17), the coefficient of the x^2-term is $1 - \lambda^2$. We know that the sum of the three roots is the negative of this coefficient. Call the third root x_3, and we have $x_1 + x_2 + x_3 = \lambda^2 - 1$, or $x_3 = \lambda^2 - 1 - x_1 - x_2$. Substitute (9.16), and this gives us:

$$x_3 = \left(\frac{(4u)(u+1)(t-1)^3 - 4t(t+1)(u-1)^3}{(4u)(u-1)(t-1)^3 - 4t(t-1)(u-1)^3} \right)^2 - 1$$

$$- \frac{4t}{(t-1)^2} - \frac{4u}{(u-1)^2}. \tag{9.18}$$

A computer algebra system now yields:

$$x_3 = \frac{4tu}{(tu-1)^2}$$

$$y_3 = -\left(\lambda(x_3 - x_1) + y_1\right) = \frac{4tu(tu+1)}{(tu-1)^3}.$$

Therefore, $\psi(tu) = (x_3, y_3)$, as promised.

Appendix D: Nonsplit Multiplicative Reduction in Detail

The formulas involved in proving (9.7) in this case are almost exactly the same as in the previous section, save for some factors involving A that do not alter the essential argument.

Assume that t, u, and tu are all different from 1, and that $t \neq u$. We compute

$$\psi(t) = \left(\frac{4At}{(t-1)^2}, \frac{4A\sqrt{A}t(t+1)}{(t-1)^3}\right) = (x_1, y_1)$$

$$\psi(u) = \left(\frac{4Au}{(u-1)^2}, \frac{4A\sqrt{A}u(u+1)}{(u-1)^3}\right) = (x_2, y_2).$$

The line through (x_1, y_1) and (x_2, y_2) has slope

$$\lambda = \frac{y_2 - y_1}{x_2 - x_1} = \frac{\frac{4A\sqrt{A}u(u+1)}{(u-1)^3} - \frac{4A\sqrt{A}t(t+1)}{(t-1)^3}}{\frac{4Au}{(u-1)^2} - \frac{4At}{(t-1)^2}}$$

$$= \sqrt{A}\,\frac{(4u)(u+1)(t-1)^3 - 4t(t+1)(u-1)^3}{(4u)(u-1)(t-1)^3 - 4t(t-1)(u-1)^3}. \tag{9.19}$$

The equation of the line through (x_1, y_1) and (x_2, y_2) is $y = \lambda(x - x_1) + y_1$. Substitute into (9.2), and we have $(\lambda(x - x_1) + y_1)^2 = x^3 + Ax^2$, or

$$0 = x^3 + Ax^2 - (\lambda(x - x_1) + y_1)^2. \tag{9.20}$$

In equation (9.20), the coefficient of the x^2-term is $A - \lambda^2$. As usual, we know that the negative of this coefficient is the sum of the three roots. Call the third root x_3, and we have $x_1 + x_2 + x_3 = \lambda^2 - A$, or $x_3 = \lambda^2 - A - x_1 - x_2$. Substitute (9.19), and this gives us:

$$x_3 = A\left(\left(\frac{(4u)(u+1)(t-1)^3 - 4t(t+1)(u-1)^3}{(4u)(u-1)(t-1)^3 - 4t(t-1)(u-1)^3} \right)^2 - 1 \right.$$
$$\left. - \frac{4t}{(t-1)^2} - \frac{4u}{(u-1)^2} \right).$$

This is exactly the same as (9.18), except for the factor of A, so we can conclude that

$$x_3 = \frac{4Atu}{(tu-1)^2}$$

$$y_3 = -\left(\lambda(x_3 - x_1) + y_1 \right) = \frac{4A\sqrt{A}tu(tu+1)}{(tu-1)^3}.$$

We can again conclude that $\psi(t) + \psi(u) = \psi(tu)$.

Chapter 10

ELLIPTIC CURVES OVER Q

Road Map

We have defined a particularly interesting finitely generated abelian group structure on the points of an elliptic curve. We now can study the torsion and rank of this group. In this way, we are beginning to close in on our quarry, counting the rational solutions to a cubic equation, or equivalently counting the rational points on an elliptic curve over **Q**.

1. The Basic Structure of the Group

In chapter 8, we took an equation of the form $y^2 = x^3 + Ax + B$, and discussed how to turn the set of solutions to this equation, along with $\mathcal{O}$, into a group. In that chapter, our examples mostly used coefficients in finite fields. That finiteness allowed us to specify the group law completely by listing all of the elements of the group. In this chapter, we turn instead to the case where the coefficients and unknowns are taken to be rational numbers. The first important theorem says that the group of rational solutions is finitely generated. It was proved in 1922 by British mathematician Louis Mordell (1888–1972).

THEOREM 10.1 (Mordell): Let E be an elliptic curve defined over **Q**. Then there exists a finite set of points $P_1, P_2, \ldots, P_n$ so that if Q is any point in $E(\mathbf{Q})$, then there are integers $m_1, m_2, m_3, \ldots, m_n$ so that $Q = m_1 P_1 + m_2 P_2 + m_3 P_3 + \cdots + m_n P_n$. In the language of chapter 7, $E(\mathbf{Q})$ is a finitely generated abelian group and the points $P_1, P_2, \ldots, P_n$ are generators of $E(\mathbf{Q})$.

Naturally, we wish to make the set of points P_k as small as possible and still allow us to write each point Q as a sum of those points. So let's suppose that we have such a "minimal set," meaning that if we deleted any one of the generators P_k from our list, we could no longer write every point in $E(\mathbf{Q})$ in terms of that shorter list.

In this situation, the generators fall into two types. Some of the points P_k will have the property that for some $a > 0$, $aP_k = \mathcal{O}$. The rest will have the property that $aP_k \neq \mathcal{O}$ for every positive integer a. The first type of generators are torsion points. Let's label those generators that are torsion points with the letters $A_1, A_2, \ldots, A_t$, and label the rest $B_1, B_2, \ldots, B_r$. It is possible for any particular elliptic curve that all of the generators might be torsion points, or some of them might be torsion points, or none of them might be torsion points. (In fact, it is possible for $E(\mathbf{Q})$ to consist of the single point $\mathcal{O}$, in which case we need no generators at all!) Because we are assuming that the set $\{A_1, \ldots, A_t, B_1, \ldots, B_r\}$ is a minimal set of generators, it is not hard to see that r is the rank of the abelian group $E(\mathbf{Q})$, as defined in chapter 7.

2. Torsion Points

We can construct the torsion subgroup of $E(\mathbf{Q})$ by taking sums of multiples of the torsion generators. Symbolically,

$$E(\mathbf{Q})_{\text{tors}} = \{a_1 A_1 + \cdots + a_t A_t \mid a_1, \ldots, a_t \text{ integers}\}.$$

This will be a finite subgroup of $E(\mathbf{Q})$. Another way to describe $E(\mathbf{Q})_{\text{tors}}$ is

$$E(\mathbf{Q})_{\text{tors}} = \{P \in E(\mathbf{Q}) \mid nP = \mathcal{O} \text{ for some positive integer } n\}.$$

This definition agrees with our definition of the torsion subgroup in chapter 7.

Surprisingly, it is possible both to compute $E(\mathbf{Q})_{\text{tors}}$ for any specific curve E, and to give a complete list of the possibilities for this subgroup. The first theorem uses the discriminant Δ_E as computed in equation (8.4), and since $\Delta_E \neq 0$, it is a powerful theorem. It was discovered separately by Norwegian mathematician Trygve Nagell (1895–1988) in 1935 and French mathematician Élisabeth Lutz (1914–2008) in 1937.

THEOREM 10.2 (Nagell–Lutz): Suppose that the equation for E has the form

$$y^2 = x^3 + a_2 x^2 + a_4 x + a_6,$$

with a_2, a_4, and a_6 integers. Suppose that $P = (x_1, y_1)$ is an element of $E(\mathbf{Q})_{\text{tors}}$. Then both x_1 and y_1 are integers, and either $y_1 = 0$, in which case $2P = \mathcal{O}$, or else y_1 divides Δ_E.

As a result of this theorem, the process of listing all of the torsion points for any particular elliptic curve E is not tricky. First, we check if setting $y = 0$ results in an equation with integer solutions; if so, then we have found some torsion points. Second, we compute Δ_E, and list all divisors (positive and negative) of Δ_E. Substitute each divisor for y, and see if the resulting equation has an integer solution for x. If it does, we then need to verify that there is some positive integer n for which $nP = \mathcal{O}$. (There is in fact an improved form of the theorem that says not just that y_1 divides Δ_E, but that y_1^2 divides Δ_E. This improvement reduces the number of possible y-coordinates of a torsion point considerably.)

The theorem also can be used to verify that a point P is not in $E(\mathbf{Q})_{\text{tors}}$. If we start with a point P, and find some positive integer n so that nP does not have integer coordinates, then we can conclude that P is not part of the torsion subgroup.

During the nineteenth century, many computations were performed on many elliptic curves. Mathematicians found different curves with points of order 2, 3, 4, 5, 6, 7, 8, 9, 10, and 12. Despite much searching no one ever was able to write down a curve E so that $E(\mathbf{Q})$ contained a point of order 11 or 13 or larger. Many people worked on proving what was and was not possible, and the culmination of that process was proved by American mathematician Barry Mazur (1937–) in 1978:

THEOREM 10.3 (Mazur): Suppose that P is a point of order n in $E(\mathbf{Q})_{\text{tors}}$. Then either $1 \leq n \leq 10$ or $n = 12$. Moreover, either

1. $E(\mathbf{Q})_{\text{tors}}$ is a cyclic group of order 1, 2, 3, 4, 5, 6, 7, 8, 9, 10, or 12. This means that

$$E(\mathbf{Q})_{\text{tors}} = \langle A_1 \rangle$$

for some point A_1 which has order 1, 2, 3, 4, 5, 6, 7, 8, 9, 10, or 12; or

2. $E(\mathbf{Q})_{\text{tors}}$ is generated by two elements A_1 and A_2, where the order of A_1 is 2, 4, 6, or 8, and the order of A_2 is 2. This means that

$$E(\mathbf{Q})_{\text{tors}} = \langle A_1, A_2 \rangle$$

where the possible orders of A_1 and A_2 are given above.

It is worth noting that the proof of theorem 10.2 takes fewer than 30 pages in Silverman and Tate (1992), while Mazur's proof of theorem 10.3 is nearly 200 pages long!

3. Points of Infinite Order

It is much harder to say anything about the rank of $E(\mathbf{Q})$. Remember that the rank is the number of points B_1, B_2, ..., B_r described in section 1 above. For a particular elliptic curve, it might be hard to compute r exactly, especially if r is larger than 7 or 8. At the time we are writing these words, the largest known rank is at least 28, for an elliptic curve found by American mathematician Noam Elkies (1966–). He has produced 28 independent nontorsion points, but it is not known if they generate the entire group $E(\mathbf{Q})$. Elkies has also found a different elliptic curve that he can prove has rank exactly 19. Here is its equation

$$y^2 + xy + y =$$

$$x^3 - x^2 + 31368015812338065133318565292206590792820353345x$$

$$+ 302038802698566087335643188429543498624522041683 87449355$$

$$5186062568159847$$

The experts expect that it should be possible to find an elliptic curve with any rank that one chooses, but a proof of this "rank conjecture" seems to be impossibly difficult at the present day.

At the same time, there are also reasons for believing that a "random" elliptic curve—that is, an equation of the form $y^2 = x^3 + Ax + B$ in which

Table 10.1. Rank and torsion of elliptic curves of the form $y^2 = x^3 + x + B$

B	Rank	Generators	$\#E(\mathbf{Q})_{\text{tors}}$
-10	0		2
-9	2	$(2, 1), (5, 11)$	1
-8	1	$(\frac{9}{4}, \frac{19}{8})$	1
-7	1	$(\frac{16}{9}, \frac{17}{27})$	1
-6	1	$(2, 2)$	1
-5	1	$(3, 5)$	1
-4	1	$(4, 8)$	1
-3	0		1
-2	0		2
-1	1	$(1, 1)$	1
0	0		2
1	1	$(0, 1)$	1
2	0		4
3	1	$(-1, 1)$	1
4	1	$(0, 2)$	1
5	0		1
6	1	$(-1, 2)$	1
7	1	$(1, 3)$	1
8	1	$(\frac{1}{4}, \frac{23}{8})$	1
9	2	$(0, 3), (8, 23)$	1
10	0		2

A and B are randomly chosen integers and $4A^3 + 27B^2 \neq 0$—should almost surely have rank either 0 or 1. The reasoning is as follows. For any given elliptic curve, it is possible (in theory at least) to predict whether its rank is even or odd. The general belief is that the rank of a random elliptic curve should also be as small as possible, given the constraint of being determined to be either even or odd (though no one has proved any statement of this form). So the experts predict that the rank should almost always be 0 or 1.

4. Examples

To show how mysteriously these quantities can vary, we conclude this chapter with table 10.1, containing the rank, a set of generators, and the size of $E(\mathbf{Q})_{\text{tors}}$ for various curves E. Rather than picking both A and B randomly, we will arbitrarily fix $A = 1$, and let B range from -10 to 10.

The computations to produce that table were done using the computer program mwrank. We leave it to you, if you wish, to find a set of generators when the rank is 0 and all the generators are torsion points.

PART III

ELLIPTIC CURVES AND ANALYSIS

Chapter 11

．． ◦ ． ◦ ．

BUILDING FUNCTIONS

Road Map

Fortified with our understanding of degree, together with some
of the algebra of elliptic curves and abelian groups, we must
now acquire more tools in order to understand the statement
of the BSD Conjecture. Our first task is to take a sequence of
numbers and use those numbers to construct functions.

1. Generating Functions

The number theory discussed in this book concerns counting the number
of solutions to systems of equations of various kinds. For example, in
Part I, we counted the number of solutions to a system of two polynomial
equations. We discussed Bézout's Theorem, which tells us how many
solutions to expect when we count properly.

In other contexts, as we will see shortly, we indulge in an infinite
sequence of counts. For example, a problem may depend on a parameter
that takes on the values 0, 1, 2, ... The result of our counting is then
an infinite sequence of numbers a_0, a_1, a_2, ... Or such a sequence may
be generated in some other way. For instance, there is the sequence 0, 1,
2, . . . itself, or even the constant sequence 1, 1, 1, ...

Mathematicians have discovered a powerful method that often helps
us to understand these sequences and predict some of their properties.
The idea is to package the sequence into a function of some kind. Then
we can use function theory to manipulate the function, and ultimately
derive hitherto unknown properties of the sequences we are studying.

Even a sequence such as 1, 1, 1, ..., suitably packaged, can help us gain information about other more complicated sequences, as we will soon see.

Let's start with any sequence of numbers $a_0, a_1, \ldots$. (Our eventual applications will typically involve a sequence of integers.) Our goal is to start with a sequence of numbers defined in one way, and somehow derive more information about the same series of numbers. Often, we would like a simple formula for a_k, but that is usually too much to hope for.

One approach is to use *generating functions*. We will use this sequence of numbers to "build" a function, which we will call $G(z)$. (The function depends on the sequence $\{a_n\}$, but that dependence is usually left out of the notation.) We use the letter z rather than x as the variable, because the variable might have to be a complex number. (It is customary to use z, w, and s as variables that take complex values.) By studying the properties of our new function $G(z)$, we hope to get insight into the sequence $\{a_n\}$ that we used to define it.

The simplest standard way to build a function is as an infinite power series. We define the function

$$G(z) = a_0 + a_1 z + a_2 z^2 + \cdots = \sum_{k=0}^{\infty} a_k z^k.$$

$G(z)$ is called a *generating function*. It gets this name from the fact that if we have some other, more compact way of writing $G(z)$, we can use that alternate formula to generate the numbers a_0, a_1, a_2, ... Here are two illustrations of how this idea can be fruitful.

The geometric series

Take the simplest example of an infinite sequence: 1, 1, 1, 1, ... The generating function in this case is:

$$G(z) = 1 + z + z^2 + z^3 + \cdots = \sum_{k=0}^{\infty} z^k.$$

We then have

$$zG(z) = z + z^2 + z^3 + \cdots = \sum_{k=1}^{\infty} z^k$$

and subtraction yields

$$G(z) - zG(z) = 1.$$

Factor out $G(z)$ from the left-hand side, and we get

$$G(z) = \frac{1}{1 - z}. \tag{11.1}$$

We've derived the formula for the sum of an infinite geometric series. We can now hint at two applications. Whenever we come up with an infinite geometric series with ratio z, we can apply our discovery above to rewrite its sum as $\frac{1}{1-z}$. This shortcut is often very useful. And we shall see in section 6 of this chapter that we can manipulate the formula for the sum of an infinite geometric series, using calculus and algebra, to obtain generating functions for important sequences of numbers that are much more complicated than $1, 1, 1, \ldots$.

Formula (11.1) cannot be used for all values of z. Substituting $z = -1$ yields

$$\frac{1}{2} \quad \text{``} = \text{''} \quad 1 - 1 + 1 - 1 + 1 \cdots,$$

a result that makes no sense. Substituting $z = 2$ yields the even stranger "equation"

$$-1 \quad \text{``} = \text{''} \quad 1 + 2 + 4 + 8 + 16 + \cdots.$$

A careful analysis shows that formula (11.1) is only valid for $|z| < 1$.

Despite this restriction, (11.1) is a *very* important formula, two of whose applications we alluded to above. In the next example, we will see how this method can be used to find an interesting formula for Fibonacci numbers.

The Fibonacci series

The Fibonacci numbers F_n are defined with the three formulas

$$F_0 = 0$$

$$F_1 = 1$$

$$F_n = F_{n-1} + F_{n-2} \qquad \text{for } n > 1. \tag{11.2}$$

The first 10 nonzero Fibonacci numbers are 1, 1, 2, 3, 5, 8, 13, 21, 34, and 55. Our goal now is to find a formula in terms of n that will give us F_n, without having to compute all of the Fibonacci numbers preceding F_n.

Let's construct a generating function using the Fibonacci numbers:

$$G(z) = 0 + z + z^2 + 2z^3 + 3z^4 + 5z^5 + \cdots = \sum_{k=0}^{\infty} F_k z^k.$$

Now we use the same trick twice, multiplying $G(z)$ by z and by z^2:

$$G(z) = z + z^2 + 2z^3 + 3z^4 + 5z^5 + 8z^6 + \cdots$$

$$zG(z) = z^2 + z^3 + 2z^4 + 3z^5 + 5z^6 + \cdots$$

$$z^2 G(z) = z^3 + z^4 + 2z^5 + 3z^6 + \cdots$$

Subtracting, we have

$$G(z) - zG(z) - z^2 G(z) = z$$

so

$$G(z) = \frac{z}{1 - z - z^2}.$$

This formula is quite helpful, but it will take a bit more algebraic work to derive the formula for F_n:

Rewrite the formula as $G(z) = \frac{-z}{z^2 + z - 1}$. Use the quadratic formula to solve the equation $z^2 + z - 1 = 0$. The two solutions are $\alpha = \frac{-1 + \sqrt{5}}{2}$ and $\beta = \frac{-1 - \sqrt{5}}{2}$. Notice that $\alpha\beta = -1$, $\alpha + \beta = -1$, and $\alpha - \beta = \sqrt{5}$. A bit of algebra then gives the equation

$$\frac{-z}{z^2 + z - 1} = \frac{1}{\sqrt{5}} \left(\frac{\alpha}{\alpha - z} - \frac{\beta}{\beta - z} \right)$$

$$= \frac{1}{\sqrt{5}} \left(\frac{1}{1 - \frac{z}{\alpha}} - \frac{1}{1 - \frac{z}{\beta}} \right).$$

Now we apply (twice) the formula for the sum of a geometric series that we derived in the previous example:

$$\frac{-z}{z^2 + z - 1} = \frac{1}{\sqrt{5}} \left(\sum_{k=0}^{\infty} \left(\frac{z}{\alpha}\right)^k - \left(\frac{z}{\beta}\right)^k \right).$$

Because the $k = 0$ term is zero, we may omit it from the sum. We end up with the equation:

$$\sum_{k=1}^{\infty} F_k z^k = \frac{1}{\sqrt{5}} \sum_{k=1}^{\infty} \left(\alpha^{-k} - \beta^{-k} \right) z^k.$$

Because $\alpha\beta = -1$, we have $\alpha^{-1} = -\beta$ and $\beta^{-1} = -\alpha$. Substitution yields:

$$\sum_{k=1}^{\infty} F_k z^k = \frac{1}{\sqrt{5}} \sum_{k=1}^{\infty} \left(\left(\frac{1 + \sqrt{5}}{2}\right)^k - \left(\frac{1 - \sqrt{5}}{2}\right)^k \right) z^k.$$

A fundamental principle of generating series is that whenever we have an equation of this sort, terms can be equated. (In order to apply this principle, it is critical that both sides of the equation must make sense for an infinite set of values of z. The reader might enjoy working out what that set of values is in this case.) As a result, we have the desired formula for the terms of the Fibonacci series:

$$F_k = \frac{1}{\sqrt{5}} \left(\left(\frac{1 + \sqrt{5}}{2}\right)^k - \left(\frac{1 - \sqrt{5}}{2}\right)^k \right). \tag{11.3}$$

If you know about the "Golden Ratio" ϕ, the appearance of $\frac{1+\sqrt{5}}{2}$ tips you off that there will be a close relationship between the Fibonacci numbers and ϕ. In the next subsection, we will learn the source of this connection.

Linear algebra and Fibonacci numbers

This section is a pure digression and can be omitted, especially if you are unfamiliar with matrices. We wish to exploit this opportunity to show how an interesting mathematical result—equation (11.3)—can be derived in more than one way. You ask: "Why bother?" After all, the preceding discussion is a convincing proof of formula (11.3). Why do mathematicians search for new proofs of a formula once they have found one proof?

One way to respond is the same way that we can answer a different question: "Why is there more than one recording of Beethoven's Fifth Symphony?" Indeed, one of the authors owns at least a half-dozen recordings of certain pieces of music. A mathematical proof is a work of art, and each new proof is another work of art, in no way redundant.

The analogy goes a bit deeper. Just as seeing multiple performances of "Hamlet" gives insights that no one performance can give, studying alternate proofs can generate new mathematical ideas. A proof of equation (11.3) using induction might not be enlightening, because an induction proof usually starts with the answer and then justifies it. However, any proof that starts with some other equation and ends with (11.3) may lead to new insights.

We use linear algebra to rewrite equation (11.2):

$$\begin{bmatrix} F_{n-1} \\ F_n \end{bmatrix} = \begin{bmatrix} 0 & 1 \\ 1 & 1 \end{bmatrix} \begin{bmatrix} F_{n-2} \\ F_{n-1} \end{bmatrix}.$$

Readers of *Fearless Symmetry* will have found a brief explanation of matrices in chapter 10. We also have the equation

$$\begin{bmatrix} F_1 \\ F_2 \end{bmatrix} = \begin{bmatrix} 0 & 1 \\ 1 & 1 \end{bmatrix} \begin{bmatrix} 0 \\ 1 \end{bmatrix}.$$

These two equations may be combined into

$$\begin{bmatrix} F_n \\ F_{n+1} \end{bmatrix} = \begin{bmatrix} 0 & 1 \\ 1 & 1 \end{bmatrix}^n \begin{bmatrix} 0 \\ 1 \end{bmatrix}. \tag{11.4}$$

Let $A = \begin{bmatrix} 0 & 1 \\ 1 & 1 \end{bmatrix}$. Let $\phi_+ = \frac{1+\sqrt{5}}{2}$ and $\phi_- = \frac{1-\sqrt{5}}{2}$. Using linear algebra, we can write $A = SBS^{-1}$, where $B = \begin{bmatrix} \phi_+ & 0 \\ 0 & \phi_- \end{bmatrix}$ and $S = \begin{bmatrix} 1 & 1 \\ \phi_+ & \phi_- \end{bmatrix}$. (This mysterious expansion comes from using the theory of eigenvalues and eigenvectors. It is not easy to stumble onto this formula if you don't know this theory, but it can easily be verified once it has been found. And if you know about eigenvalues, this formula would occur to you naturally.) The formula $\begin{bmatrix} a & b \\ c & d \end{bmatrix}^{-1} = \frac{1}{ad-bc} \begin{bmatrix} d & -b \\ -c & a \end{bmatrix}$ lets us write $S^{-1} = \frac{-1}{\sqrt{5}} \begin{bmatrix} \phi_- & -1 \\ -\phi_+ & 1 \end{bmatrix}$. The value of the equation $A = SBS^{-1}$ is that if n is any integer, we have $A^n = SB^n S^{-1}$. Moreover, while there is no obvious formula for A^n, it is easy to see that $B^n = \begin{bmatrix} \phi_+^n & 0 \\ 0 & \phi_-^n \end{bmatrix}$.

All of this algebra combines to tell us that (11.4) becomes

$$\begin{bmatrix} F_n \\ F_{n+1} \end{bmatrix} = S \begin{bmatrix} \phi_+^n & 0 \\ 0 & \phi_-^n \end{bmatrix} S^{-1} \begin{bmatrix} 0 \\ 1 \end{bmatrix}$$

$$= \begin{bmatrix} 1 & 1 \\ \phi_+ & \phi_- \end{bmatrix} \begin{bmatrix} \phi_+^n & 0 \\ 0 & \phi_-^n \end{bmatrix} \left(\frac{-1}{\sqrt{5}} \right) \begin{bmatrix} \phi_- & -1 \\ -\phi_+ & 1 \end{bmatrix} \begin{bmatrix} 0 \\ 1 \end{bmatrix}$$

$$= \frac{-1}{\sqrt{5}} \begin{bmatrix} 1 & 1 \\ \phi_+ & \phi_- \end{bmatrix} \begin{bmatrix} \phi_+^n & 0 \\ 0 & \phi_-^n \end{bmatrix} \begin{bmatrix} -1 \\ 1 \end{bmatrix}$$

$$= \frac{-1}{\sqrt{5}} \begin{bmatrix} 1 & 1 \\ \phi_+ & \phi_- \end{bmatrix} \begin{bmatrix} -\phi_+^n \\ \phi_-^n \end{bmatrix}$$

$$= \frac{-1}{\sqrt{5}} \begin{bmatrix} -\phi_+^n + \phi_-^n \\ -\phi_+^{n+1} + \phi_-^{n+1} \end{bmatrix} = \frac{1}{\sqrt{5}} \begin{bmatrix} \phi_+^n - \phi_-^n \\ \phi_+^{n+1} - \phi_-^{n+1} \end{bmatrix}.$$

We can now easily read off the equation

$$F_n = \frac{1}{\sqrt{5}} \left(\phi_+^n - \phi_-^n \right),$$

which is our new derivation of (11.3).

Notice that in this derivation, the numbers ϕ_+ and ϕ_- appear naturally as the eigenvalues of the matrix A. The number ϕ_+ is called the "Golden Ratio." You can read about it in many books, including (Livio, 2002).

2. Dirichlet Series

Although generating functions give us a lot of information, they are not the only way to go. We can start with a sequence of numbers $a_0, a_1, \ldots,$ and produce a type of function first defined by German mathematician Johann Peter Gustav Lejeune Dirichlet (1805–59).

DEFINITION: A *Dirichlet series* is a function of the form

$$\sum_{n=1}^{\infty} \frac{a_n}{n^s}.$$

We often will write this sum as $\sum a_n n^{-s}$, where we implicitly sum using the variable n.

Notice that this sum starts at $n = 1$, so a_0 is not relevant. Moreover, in many applications, we require that $a_1 = 1$. A famous example of a Dirichlet series is the Riemann zeta-function.

We will need to substitute complex values for the variable s, so we remind you how to define n^s when n is a positive integer and s is complex. Remember first that $n = e^{\log n}$. (Here we follow the convention of all mathematicians who are not teaching freshman calculus and write $\log n$ to mean the natural logarithm, that is, the logarithm to the base e, of the variable n.) If $s = \sigma + it$, where σ and t are real numbers, then $n^s = e^{s \log n} = e^{(\sigma + it) \log n} = e^{\sigma \log n} e^{i(t \log n)}$. Now, there is no trouble computing $e^{\sigma \log n}$, because σ and $\log n$ are just real numbers. For $e^{i(t \log n)}$, we use the formula $e^{i\theta} = \cos \theta + i \sin \theta$, and so $e^{i(t \log n)} = \cos(t \log n) + i \sin(t \log n)$. For example, $2^{3+4i} = e^{(3+4i) \log 2} = e^{3 \log 2} e^{i(4 \log 2)} = 8 e^{i(4 \log 2)} = 8(\cos(4 \log 2) + i \sin(4 \log 2)) \approx 8(-0.9327 + 0.3607i) \approx -7.4615 + 2.8855i$. (By the way, the notation $s = \sigma + it$ has been standard for over a century.)

Because we will be substituting complex numbers and adding up Dirichlet series, we need to discuss when this is possible. The terminology we use is the usual one from calculus.

DEFINITION: If

$$\lim_{m \to \infty} \sum_{n=1}^{m} \frac{a_n}{n^s}$$

exists and is equal to the number Z, then the Dirichlet series *converges* to Z.

The most basic result about convergent Dirichlet series is the following.

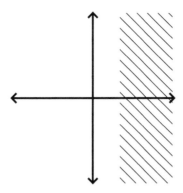

Figure 11.1. The region $\sigma > 1$

THEOREM 11.5: Suppose that there is some constant K so that the numbers $|a_n| < K$ for all n. (We say in this case that the numbers a_n are *bounded*.) Then the Dirichlet series $\sum a_n n^{-s}$ converges if $\sigma > 1$.

If you learned about absolutely convergent series in your calculus class, it would not be too hard for you to prove theorem 11.5, but we will take the theorem as given. The set of complex numbers s with $\sigma > 1$ is called a *complex half-plane*, and is depicted in figure 11.1.

A bit more is true, and we record that fact as well.

THEOREM 11.6: Suppose that there is some constant K so that $|a_n| < Kn^r$ for all n. Then the Dirichlet series $\sum a_n n^{-s}$ converges if $\sigma > r + 1$.

We need to state yet one more result about when a Dirichlet series converges:

THEOREM 11.7: Suppose that there is some constant K so that $|a_1 + \cdots + a_n| < K$ for all n. Then the Dirichlet series $\sum a_n n^{-s}$ converges if $\sigma > 0$.

3. The Riemann Zeta-Function

The simplest case of a Dirichlet series occurs when we set $a_n = 1$ for all n. The result is the *Riemann zeta-function*:

$$\zeta(s) = 1 + \frac{1}{2^s} + \frac{1}{3^s} + \frac{1}{4^s} + \frac{1}{5^s} + \frac{1}{6^s} + \frac{1}{7^s} + \frac{1}{8^s} + \cdots .$$

Bernhard Riemann was a German mathematician who lived from 1826 to 1866, and his famous "Riemann Hypothesis" is a conjecture about when $\zeta(s)$ equals zero.

Theorem 11.5 tells us that this series converges if $\sigma > 1$. Substitution of $s = 1$ produces the harmonic series

$$1 + \frac{1}{2} + \frac{1}{3} + \frac{1}{4} + \frac{1}{5} + \cdots,$$

which is well known to be divergent. Our goal is to find other expressions for the zeta-function so that it can be evaluated somehow when $\sigma \leq 1$. This is the process of analytic continuation we will discuss in section 4 of chapter 12, but in this case we want to avoid using Taylor series. Taylor series are useful for explaining the abstract theory of analytic continuation and for proving theorems about it, but they are often not too helpful in actually computing the analytic continuation of specific functions.

We start with a trick. Multiply the sum for $\zeta(s)$ by $1/2^s$, and we get:

$$\frac{1}{2^s}\zeta(s) = \frac{1}{2^s} + \frac{1}{4^s} + \frac{1}{6^s} + \frac{1}{8^s} + \cdots.$$

Let's line this up—the trick is to do it twice—underneath the sum for $\zeta(s)$, and subtract:

$$\zeta(s) = 1 + \frac{1}{2^s} + \frac{1}{3^s} + \frac{1}{4^s} + \frac{1}{5^s} + \frac{1}{6^s} + \frac{1}{7^s} + \frac{1}{8^s} + \frac{1}{9^s} + \frac{1}{10^s} + \cdots$$

$$\frac{1}{2^s}\zeta(s) = \quad \frac{1}{2^s} \quad + \frac{1}{4^s} \quad + \frac{1}{6^s} \quad + \frac{1}{8^s} \quad + \frac{1}{10^s} + \cdots$$

$$\frac{1}{2^s}\zeta(s) = \quad \frac{1}{2^s} \quad + \frac{1}{4^s} \quad + \frac{1}{6^s} \quad + \frac{1}{8^s} \quad + \frac{1}{10^s} + \cdots$$

$$\zeta(s) - \frac{1}{2^s}\zeta(s) - \frac{1}{2^s}\zeta(s) = 1 - \frac{1}{2^s} + \frac{1}{3^s} - \frac{1}{4^s} + \frac{1}{5^s} - \frac{1}{6^s} + \frac{1}{7^s} - \frac{1}{8^s} + \frac{1}{9^s} - \frac{1}{10^s} + \cdots$$

The result is

$$\left(1 - \frac{1}{2^{s-1}}\right)\zeta(s) = 1 - \frac{1}{2^s} + \frac{1}{3^s} - \frac{1}{4^s} + \frac{1}{5^s} - \frac{1}{6^s} + \frac{1}{7^s} - \frac{1}{8^s} + \frac{1}{9^s} - \frac{1}{10^s} + \cdots.$$

$$(11.8)$$

In equation (11.8), the right-hand side is a Dirichlet series in which the coefficients are $1, -1, 1, -1, 1, -1, \ldots$. Notice that $|a_1 + a_2 + \cdots + a_n| < 2$ for any value of n. Theorem 11.7 now tells us that the right-hand side of equation (11.8) can be summed provided that $\sigma > 0$. The formula

$$\zeta(s) = \left(1 - \frac{1}{2^{s-1}}\right)^{-1}\left(1 - \frac{1}{2^s} + \frac{1}{3^s} - \frac{1}{4^s} + \frac{1}{5^s} - \frac{1}{6^s}\right.$$

$$\left. + \frac{1}{7^s} - \frac{1}{8^s} + \frac{1}{9^s} - \frac{1}{10^s} + \cdots\right)$$

therefore can be evaluated provided that $\sigma > 0$, with the sole exception of the value $s = 1$.

However, we need to do still more, and find a way to evaluate $\zeta(s)$ for *all* values of s except $s = 1$. That requires a discussion of *functional equations*.

4. Functional Equations

Let's take not one but two steps back, and briefly discuss elementary trigonometric functions. When the sine and cosine functions are first discussed in algebra classes, they are defined using a right triangle with an acute angle x. The definitions given are

$$\sin x = \frac{\text{opposite}}{\text{hypotenuse}} \qquad \cos x = \frac{\text{adjacent}}{\text{hypotenuse}}.$$

The definitions only make sense if $0 \le x \le \frac{\pi}{2}$. There are several ways to extend the definition of these functions to any value of x. One way is to make use of the formulas

$$\sin(x + y) = \sin x \cos y + \cos x \sin y$$
$$\cos(x + y) = \cos x \cos y - \sin x \sin y.$$
(11.9)

The first formula allows us to compute that $\sin(x + \frac{\pi}{2}) = \cos x$, while the second tells us that $\cos(x + \frac{\pi}{2}) = -\sin x$. This enables us to compute $\sin x$ and $\cos x$ for $0 \le x \le \pi$. But then we can apply the formulas again to see that $\sin(x + \pi) = -\sin x$ and $\cos(x + \pi) = -\cos x$, enabling us to compute $\sin x$ and $\cos x$ for $0 \le x \le 2\pi$. And finally the formulas $\sin(x + 2\pi) = \sin x$ and $\cos(x + 2\pi) = \cos x$ can be applied repeatedly to evaluate $\sin x$ and $\cos x$ for all real numbers x.

Note that by starting with definitions of $\sin x$ and $\cos x$ for a limited set of x-values, along with an equation relating $\sin x$ and $\cos x$ to other values of the sine and cosine functions, we ended up with definitions for all x-values.

DEFINITION: A *functional equation* is an equation that relates the value of some function $f(x)$ to the same function evaluated at some number other than x.

The equations $\sin(x + \pi) = -\sin x$ and $\sin(x + 2\pi) = \sin x$ are both examples of functional equations.

If $h(t) = f(t) + ig(t)$ is a complex-valued function of a real variable t, then we define

$$\int h(t)\, dt = \int f(t)\, dt + i \int g(t)\, dt.$$

We make a step back toward the Riemann zeta-function, and define the gamma-function:

DEFINITION: The *gamma-function* is defined by the equation

$$\Gamma(z) = \int_0^\infty e^{-t} t^{z-1}\, dt \tag{11.10}$$

if z is any complex number whose real part is positive.

We will not go through the details, but equation (11.10) in fact does yield a convergent integral if $z = x + iy$ and $x > 0$. The gamma-function was originally defined in an attempt to expand the definition of factorial to noninteger values; its history is beautifully summarized in (Davis, 1959). Recall the definition of factorial: $0! = 1$, $1! = 1$, and if $n > 1$, $n! = 1 \cdot 2 \cdot 3 \cdots n$.

The key properties of $\Gamma(z)$ are:

$$\Gamma(1) = 1$$

$$\Gamma(z + 1) = z\Gamma(z)$$

$$\Gamma(n + 1) = n! \quad \text{if } n \text{ is a nonnegative integer.}$$

Notice that the equation $\Gamma(z+1) = z\Gamma(z)$ is a functional equation, because it relates the values $\Gamma(z)$ and $\Gamma(z+1)$.

With some effort, it can be shown that the product $\sin(\pi z)\Gamma(z)$ is well-defined for all complex numbers z, and therefore $\Gamma(z)$ is defined except possibly when $\sin(\pi z)$ is 0. In particular, the formula $\Gamma(z+1) = z\Gamma(z)$ can be rewritten in the form $\Gamma(z) = \Gamma(z+1)/z$, allowing us to compute $\Gamma(z)$ for any value of z other than a negative integer or 0. (In fact, $\Gamma(z)$ has a *pole of order* 1 at $z = k$ for every nonpositive integer k. See page 198 for the definition of "pole of order 1.") Let's suppose we wanted to compute $\Gamma(-2+i)$. We can start with equation (11.10), and use numerical integration to approximate $\Gamma(1+i)$. We get $\Gamma(1+i) \approx 0.498 - 0.155i$. Therefore, $\Gamma(i) = \Gamma(1+i)/i \approx -0.155 - 0.498i$. Then $\Gamma(-1+i) = \Gamma(i)/(-1+i) \approx -0.172 + 0.326i$. Finally, $\Gamma(-2+i) = \Gamma(-1+i)/(-2+i) \approx 0.134 - 0.096i$.

With a lot of effort, it can be shown that

$$\Gamma(z)\Gamma(1-z) = \frac{\pi}{\sin \pi z}$$

which shows that $\Gamma(z)$ has a serious relationship to the more familiar sine function.

We return, at long last, to $\zeta(s)$. It is possible to prove that the function $(s-1)\zeta(s)$ is defined for all complex values of s, and therefore that $\zeta(s)$ is defined except when $s = 1$. Moreover, $\zeta(s)$ satisfies the functional equation:

$$\zeta(1-s) = 2^{1-s}\pi^{-s}\cos\left(\frac{\pi s}{2}\right)\Gamma(s)\zeta(s). \qquad (11.11)$$

Equation (11.11) is sufficiently interesting that there are several proofs of it in (Titchmarsh, 1986), the classic reference about the zeta-function. In particular, equation (11.11) allows us to compute $\zeta(s)$ for all values of s. For example, if we want to compute $\zeta(-1.8)$, we can substitute $s = -1.8$ into the equation, and get

$$\zeta(2.8) = 2^{2.8}\pi^{1.8}\cos(-0.9\pi)\Gamma(-1.8)\zeta(-1.8)$$

We can approximate $\zeta(2.8)$ using the definition of $\zeta(s)$, and we get $\zeta(2.8) \approx 1.247$. To approximate $\Gamma(-1.8)$, we start with $\Gamma(2.2) \approx 1.102$, and then $\Gamma(1.2) \approx 0.918$, $\Gamma(0.2) \approx 4.591$, $\Gamma(-0.8) \approx -5.739$, and $\Gamma(-1.8) \approx 3.188$. Substitution of these values tells us that $\zeta(-1.8) \approx -0.00752$.

5. Euler Products

We are not done exploiting various properties of $\zeta(s)$. Notice that we can factor the defining formula for $\zeta(s)$, using unique prime factorization of integers:

$$\zeta(s) = 1 + \frac{1}{2^s} + \frac{1}{3^s} + \frac{1}{4^s} + \frac{1}{5^s} + \frac{1}{6^s} + \frac{1}{7^s} + \frac{1}{8^s} + \frac{1}{9^s} + \frac{1}{10^s} + \cdots \qquad (11.12)$$

$$= \left(1 + \frac{1}{2^s} + \frac{1}{4^s} + \frac{1}{8^s} + \frac{1}{16^s} + \cdots\right) \times \left(1 + \frac{1}{3^s} + \frac{1}{9^s} + \frac{1}{27^s} + \frac{1}{81^s} + \cdots\right)$$

$$\times \left(1 + \frac{1}{5^s} + \frac{1}{25^s} + \frac{1}{125^s} + \frac{1}{625^s} + \cdots\right)$$

$$\times \left(1 + \frac{1}{7^s} + \frac{1}{49^s} + \frac{1}{343^s} + \frac{1}{2401^s} + \cdots\right) \times \cdots .$$

Every integer n in the sum can be factored uniquely as a product of powers of distinct primes, and we use that factorization to split the sum into an infinite product of infinite sums. (We'll see in a moment why writing a single infinite sum as an infinite product of infinite sums consists of progress rather than madness.) The notation gets awful because writing a factorization of n is unpleasant, but here goes. If $n = p_1^{e_1} p_2^{e_2} \cdots p_k^{e_k}$, then

$$\frac{1}{n^s} = \left(\frac{1}{p_1^{e_1 s}}\right)\left(\frac{1}{p_2^{e_2 s}}\right) \cdots \left(\frac{1}{p_k^{e_k s}}\right).$$

We regroup the terms, and get equation (11.12), which we can now rewrite as 2 equations:

$$\zeta(s) = \prod_p \zeta_p(s)$$

$$\zeta_p(s) = 1 + \frac{1}{p^s} + \frac{1}{p^{2s}} + \frac{1}{p^{3s}} + \frac{1}{p^{4s}} + \cdots .$$

(In the first equation, the notation $\prod_p$ means that we take an infinite product where the variable p ranges over all primes.) The key observation, which makes all of our work enlightening rather than confusing, is that $\zeta_p(s)$ is a geometric series with ratio $1/p^s$. Therefore, equation (11.1) says that

$$\zeta_p(s) = 1 + \frac{1}{p^s} + \frac{1}{p^{2s}} + \frac{1}{p^{3s}} + \cdots = \frac{1}{1 - \frac{1}{p^s}}.$$

In order to write terms a bit more compactly, we write $\zeta_p(s) = (1 - \frac{1}{p^s})^{-1}$, or even $\zeta_p(s) = (1 - p^{-s})^{-1}$.

Put everything together, and we have the wonderful formula

$$\zeta(s) = \prod_p \left(1 - \frac{1}{p^s}\right)^{-1}. \tag{11.13}$$

Equation (11.13) is the easiest way to understand why information about $\zeta(s)$ can be exploited to give information about prime numbers.

We will not go any further down that road. Equation (11.13) is also an example of an *Euler product*. An Euler product is an infinite product of simple functions of prime numbers, and we will see more of them in chapter 13.

How much of what we did above can be repeated if we start with an arbitrary Dirichlet series? Suppose that the coefficients a_m satisfy the property that $a_m a_n = a_{mn}$ whenever the integers m and n have no common divisor other than 1. For example, $a_8 a_{25} = a_{200}$. (On the other hand, we do not ask for $a_8 a_{12}$ to be related in any way to a_{96}, for example.) In this case, we can repeat the above factorization. Suppose that $L(s) = \sum a_n n^s$. We can compute

$$L(s) = 1 + \frac{a_2}{2^s} + \frac{a_3}{3^s} + \frac{a_4}{4^s} + \frac{a_5}{5^s} + \frac{a_6}{6^s} + \frac{a_7}{7^s} + \frac{a_8}{8^s} + \frac{a_9}{9^s} + \frac{a_{10}}{10^s} + \cdots \tag{11.14}$$

$$= \left(1 + \frac{a_2}{2^s} + \frac{a_4}{4^s} + \frac{a_8}{8^s} + \frac{a_{16}}{16^s} + \cdots\right)$$

$$\times \left(1 + \frac{a_3}{3^s} + \frac{a_9}{9^s} + \frac{a_{27}}{27^s} + \frac{a_{81}}{81^s} + \cdots\right)$$

$$\times \left(1 + \frac{a_5}{5^s} + \frac{a_{25}}{25^s} + \frac{a_{125}}{125^s} + \frac{a_{625}}{625^s} + \cdots\right)$$

$$\times \left(1 + \frac{a_7}{7^s} + \frac{a_{49}}{49^s} + \frac{a_{343}}{343^s} + \frac{a_{2401}}{2401^s} + \cdots\right) \times \cdots$$

Just as before, we can rewrite equation (11.14)

$$L(s) = \prod_p L_p(s)$$

$$L_p(s) = \sum_{k=0}^{\infty} \frac{a_{p^k}}{p^{ks}}.$$

The simplification process might or might not end here. In the examples that we will consider later in this book, the coefficient a_p will determine a_{p^k} for $k > 1$, and then it will be possible to write the functions $L_p(s)$ in short formulas that only contain the number a_p. The numbers a_p will be defined in terms of the number of solutions to a cubic equation modulo p.

In fact, we will typically proceed in the reverse direction. Rather than define a_n for every integer n, we will give a formula for $L_p(s)$, and then we will define

$$L(s) = \prod_p L_p(s) = \sum \frac{a_n}{n^s}.$$

In principle, this formula allows the computation of a_n for every integer n, but most of the interesting information (for us) is already contained in the numbers a_p.

6. Build Your Own Zeta-Function

Of all of the methods for constructing functions in this chapter, this last one is the method most important for elliptic curves. We begin with a sequence of numbers $M_1, M_2, M_3, \ldots$. In our application, these numbers will count solutions to systems of equations and so will be positive integers, but the construction works in general.

We construct a function $Z(T)$, which depends on the integers M_r, according to the formula

$$Z(T) = \exp\left(\sum_{r=1}^{\infty} \frac{M_r T^r}{r}\right). \tag{11.15}$$

Here we use the notation

$$\exp(x) = e^x = 1 + x + \frac{x^2}{2!} + \frac{x^3}{3!} + \cdots$$

The importance of $Z(T)$ is that it packages the numbers M_r in such a way that we can use $Z(T)$ to study the numbers M_r, and thereby to study whatever system of equations gave rise to the numbers M_r.

It is always a good idea to explore extreme cases. For this function, there is one extreme case which is easy to understand. If the numbers M_r are all 0, then we get

$$Z(T) = \exp\left(\sum_{r=1}^{\infty} \frac{0 \cdot T^r}{r}\right) = \exp(0) = 1.$$

That example didn't teach us all that much, other than that the formula might not be as daunting to work with as it may seem at first.

We can also try $M_1 = 1$, and $M_r = 0$ if $r \neq 1$. In that case, we get

$$Z(T) = \exp\left(\sum_{r=1}^{\infty} \frac{M_r T^r}{r}\right) = \exp(T),$$

which is also not too terrifying.

Next, we have to try something slightly harder. What happens if $M_r = 1$ for all r? We get

$$Z(T) = \exp\left(\sum_{r=1}^{\infty} \frac{T^r}{r}\right)$$

which does not look too approachable. We do the only thing that we can do, and try to unpack some of the complexity by taking the natural logarithm of both sides of the equation. We get

$$\log Z(T) = \sum_{r=1}^{\infty} \frac{T^r}{r}$$

which looks just as bad. Now, we take derivatives, and some amazing things happen. The chain rule from calculus tells us that $\frac{d}{dt} \log Z(T) = \frac{Z'(T)}{Z(T)}$, while we can differentiate the sum on the right-hand side of the equation one term at a time in "dummy" fashion. We get

$$\frac{Z'(T)}{Z(T)} = \sum_{r=1}^{\infty} T^{r-1} = 1 + T + T^2 + T^3 + \cdots.$$

But now we can recognize the right-hand side of this last equation as a geometric series, and use the very first formula of this chapter to get

$$\frac{Z'(T)}{Z(T)} = \frac{1}{1-T}.$$

Now, we can reverse the process and integrate both sides of the equation. We get

$$\log Z(T) = \int \frac{1}{1-T} \, dT = -\log(1-T) = \log\left(\frac{1}{1-T}\right)$$

and so $Z(T) = \frac{1}{1-T}$. (If you are worried about a constant of integration, a possibility that we ignored in our derivation, you can substitute $T = 0$ and see that in fact there is no need to worry.) We can record all of this in a single equation:

$$\exp\left(\sum_{r=1}^{\infty} \frac{T^r}{r}\right) = \frac{1}{1-T}. \tag{11.16}$$

It turns out that equation (11.16) is the main tool that we need to understand the basic properties of the function $Z(T)$. Start by replacing T by αT in equation (11.16), and we get

$$\exp\left(\sum_{r=1}^{\infty} \frac{\alpha^r T^r}{r}\right) = \frac{1}{1-\alpha T}. \tag{11.17}$$

Equation (11.17) tells us that if $M_r = \alpha^r$, then $Z(T) = \frac{1}{1-\alpha T}$. Suppose now that we take the reciprocal of both sides of equation (11.17). We get

$$\frac{1}{\exp\left(\sum_{r=1}^{\infty} \frac{\alpha^r T^r}{r}\right)} = 1 - \alpha T. \tag{11.18}$$

Equation (11.18) looks as if it is not good for much until we recall that $\frac{1}{e^x} = e^{-x}$, and then we can rewrite equation (11.18) as

$$\exp\left(\sum_{r=1}^{\infty} \frac{-\alpha^r T^r}{r}\right) = 1 - \alpha T. \tag{11.19}$$

Equation (11.19) tells us that if $M_r = -\alpha^r$, then $Z(T) = 1 - \alpha T$.

There is only one more observation to make, and we'll phrase it as an exercise.

EXERCISE: Suppose that we start with two sequences of numbers M_r and N_r, and define

$$Z_1(T) = \exp\left(\sum_{r=1}^{\infty} \frac{M_r T^r}{r}\right)$$

$$Z_2(T) = \exp\left(\sum_{r=1}^{\infty} \frac{N_r T^r}{r}\right)$$

$$Z_3(T) = \exp\left(\sum_{r=1}^{\infty} \frac{(M_r + N_r)T^r}{r}\right).$$

Show that $Z_1(T)Z_2(T) = Z_3(T)$.

SOLUTION: This exercise is just a slightly disguised application of the formula $\exp(a + b) = \exp(a)\exp(b)$.

The exercise tells us that

$$\exp\left(\sum_{r=1}^{\infty} \frac{(\alpha^r - \beta^r)T^r}{r}\right) = \frac{1 - \beta T}{1 - \alpha T}.$$

Building on this idea, we can summarize this section with the following theorem.

THEOREM 11.20: Suppose that we have a sequence of numbers M_r, and use these numbers to define a function $Z(T)$ with the formula

$$Z(T) = \exp\left(\sum_{r=1}^{\infty} \frac{M_r T^r}{r}\right).$$

Suppose that we can show that

$$Z(T) = \frac{f(T)}{g(T)}$$

where $f(T)$ and $g(T)$ are polynomials that factor as

$$f(T) = (1 - \beta_1 T)(1 - \beta_2 T) \cdots (1 - \beta_j T)$$

$$g(T) = (1 - \alpha_1 T)(1 - \alpha_2 T) \cdots (1 - \alpha_k T).$$

Then for every positive integer r, $M_r = \alpha_1^r + \alpha_2^r + \cdots + \alpha_k^r - \beta_1^r - \beta_2^r - \cdots - \beta_j^r$.

Theorem 11.20 will play a crucial role in chapter 13. In that chapter, we will set the numbers M_r equal to the number of solutions to some polynomial equation, where we draw the values of the variables from the finite fields $\mathbf{F}_{p^r}$. Then theorem 11.20 tells us that the infinite sequence M_1, M_2, M_3, ... is determined by a finite amount of data $\alpha_1, \ldots, \alpha_k, \beta_1, \ldots, \beta_j$, and those α's and β's are key numbers in understanding the equation that we started with.

Chapter 12

ANALYTIC CONTINUATION

Road Map

To count points on elliptic curves efficiently and to formulate the Birch–Swinnerton-Dyer Conjecture, we will package
certain counts in the form of *L*-functions, using ideas from
chapter 11. Then we will have to study the properties of the
L-functions. They are functions of a complex variable that
are differentiable in a certain strong sense, earning them the
epithet "analytic." In this chapter, we will explain what analytic
functions are and explore some of their properties.

Every polynomial is "analytic." So we begin by discussing
polynomials and how all of their values can be determined if
you know enough of their values. We will then see that analytic
functions have a similar property, which enables us to extend
the domain of a given analytic function. This is related to the
fact that we could extend the domains of the gamma-function
and the Riemann zeta-function in chapter 11. Later, crucially,
we will extend the domains of *L*-functions of elliptic curves.

1. A Difference that Makes a Difference

Think of a polynomial in one variable—we will use $f(x) = 2x^2 - 4x + 1$ as
an example—and make a table as follows (see table 12.1): List a few entries
for consecutive integer values of x. Underneath the first row, containing
the values of $f(x)$ for those consecutive integers x, list the row of successive
differences. For example, under the -1 in the first row, we list -2 because

Table 12.1. Successive differences

$f(0)$	$f(1)$	$f(2)$	$f(3)$	$f(4)$	...	$f(100)$
1	−1	1	7	?	...	?
2	−2	−6	?	...	...	?
4	4	?	?	...	...	?
0	?	...	...	...	...	?

Table 12.2. Successive differences reversed: step 1

$f(0)$	$f(1)$	$f(2)$	$f(3)$	$f(4)$	...	$f(100)$
1	−1	1	7	?	...	?
2	−2	−6	?	...	...	?
4	4	4	4	...	...	4
0	0	...	...	...	...	0

that is the difference of −1 and the following value 1. Now repeat this process again on the second row, and then the third row, and so on.

We may as well stop when we get a row of zeros, since every row thereafter will also be all zeros. It turns out that, no matter what polynomial you start with, you will *always* get to a row of all zeros eventually. It doesn't depend which integer value you start with (in our example we started with $x = 0$) nor with how far out you extend the row. In fact, you can prove that if you start with a polynomial of degree d, then the $(d + 2)$-nd row will be all zeros. In our example, the fourth row will consist solely of zeros.

This fact—we could call it the "zeroing out of successive differences" for polynomials—has a far-reaching significance. Suppose you had only table 12.1, without any knowledge of the identity of the polynomial $f(x)$. Suppose you even don't know how to multiply numbers. You can still figure out what $f(100)$ is, once you are told that the bottom row is all zeros, infinitely far to the right and the left (which is true). How? First, extend the bottom row of zeros all the way to the right to the $f(100)$ column. Then, build up the next row above by subtracting the value directly below you from the value where you are, and enter it to the right of where you are. For example, in table 12.1, the "?" after the second 4 in the third row would be replaced by $4 − 0 = 4$. This is because we know that if we do the usual subtraction on the third row, we must get back the fourth row. After completing the row of 4's, our table looks like table 12.2.

We continue this process by filling in the second row in the same way. After each row is complete, we go to the row above. For example, in the

Table 12.3. Successive differences reversed: steps 2 and 3

$f(0)$	$f(1)$	$f(2)$	$f(3)$	$f(4)$	...	$f(100)$
1	-1	1	7	17	...	19601
2	-2	-6	-10	...	...	?
4	4	4	4	...	...	4
0	0	...	...	...	...	0

second row next to the -6, we must enter $-6 - 4 = -10$. Filling in the rest of the entries this way, we get table 12.3.

We can now see that $f(4)$ is predicted to be $7 - (-10) = 17$, and indeed it is. It would be rather laborious to carry this all the way out to determine $f(100)$, but it could be done. And it wouldn't even take too long if you used a computer. So you can find $f(100)$ from the knowledge that $f(x)$ has degree 2 and from a *single* column of numbers, for example the column 1, 2, 4, and 0 below $f(0)$. You only need to know a single number in a row to compute the entire row, provided that you know all the numbers in the row below it.

This procedure is usually not an efficient way to compute the values of polynomials, but it makes a very interesting theoretical point. It says that polynomials have a powerful property of "action at a distance." If $f(x)$ is degree d and you know the values of $f(x)$ for $d + 1$ consecutive integers, you can use subtraction to find the column of numbers below the first integer. Then you can use other subtractions to fill in all the rows from the bottom up, starting with a row of all zeroes. In this way, you can find the values of $f(x)$ for *any* integer value of x by reconstructing the table, no matter how far away x is from the values you know. Don't forget that we know *a priori* that the $(d + 2)$-nd row is all zeros, so we can fill that row in first.

You may know of the following theorem: The coefficients of a polynomial $f(x)$ of degree d (and therefore the values of $f(x)$ for all x) are determined by the values of $f(a_0), \ldots, f(a_d)$ for any $d + 1$ numbers $a_0, \ldots, a_d$, as long as they are $d + 1$ distinct numbers. This is done by using *Lagrange interpolation*, and is often covered in first-year calculus classes. (Joseph-Louis Lagrange was an Italian–French mathematician who lived from 1736 to 1816.) Lagrange interpolation requires multiplication and division as well as addition and subtraction; in other words, Lagrange interpolation requires a field.

The theorem about Lagrange interpolation makes sense. A polynomial of degree d is determined by $d + 1$ numbers, namely its coefficients. So if you are given the $d + 1$ data points $(a_0, f(a_0)), \ldots, (a_d, f(a_d))$, it is not surprising that you can find the $d + 1$ coefficients of $f(x)$. Not surprising, but it does require proof.

The concept of Lagrange interpolation can be generalized in various ways. For example, if you have too many data points, many more than the degree d of the unknown polynomial, you can "fit the curve" to the data. That is, you can try to find a polynomial of degree d that does the "best" job of going through or near the given data points. There are various definitions of "best" depending on your purposes, which are discussed in the area of mathematics called *numerical analysis*. Statisticians use *curve fitting* extensively.

There is another set of "data" that can be used to find f. We computed successive differences to find table 12.1. The successive difference at the integer a is defined to be $f(a + 1) - f(a)$, and we can write that suggestively as:

$$f(a + 1) - f(a) = \frac{f(a + 1) - f(a)}{(a + 1) - a}. \tag{12.1}$$

The right-hand side of equation 12.1 is supposed to suggest the difference quotient used to define the derivative of $f(x)$. Instead of using these successive discrete difference quotients, let's use the derivative itself.

For example, using the same $f(x) = 2x^2 - 4x + 1$, compute the successive derivatives of $f(x)$ at 0. The "zeroth" derivative by definition is $f^{(0)}(x) = f(x) = 2x^2 - 4x + 1$. The first derivative is $f^{(1)}(x) = f'(x) = 4x - 4$. The second derivative is $f^{(2)}(x) = f''(x) = 4$, and the third and all higher derivatives are 0. Again, you can see the role of degree. If you start with a polynomial of degree d, its $(d + 1)$-st and all higher derivatives are 0.

Notice the notation we will use for derivatives. The n-th derivative of $f(x)$ is $f^{(n)}(x)$. This notation is handy for large values of n, for symbolic purposes where we don't want to specify n, and for summation purposes when we want to sum over n.

Now evaluate these derivatives at 0. You get $f^{(0)}(0) = 1$, $f^{(1)}(0) = -4$, and $f^{(2)}(0) = 4$. With the knowledge that $f(x)$ has degree 2, we can reconstruct $f(x)$ from these data points alone, using "Taylor series." This is another manifestation of the "action at a distance" of polynomials. Here we only need to know some of the derivatives of a polynomial at a *single* point, and we can figure out what $f(x)$ is—and therefore what its values are anywhere—no matter how far we travel from the x-value we started with. To explain this concept further, we need to review Taylor series briefly.

2. Taylor Made

Suppose $f(x)$ is a function from an open interval U to $\mathbf{R}$, and suppose a is a number in the interval U. Assume that $f(x)$ and all its derivatives are defined at a. Recall the definition of $n!$ from chapter 11, section 4.

> **DEFINITION**: The *Taylor series of f at a*, denoted $\text{Taylor}_a(f)$, is defined to be the infinite sum
>
> $$\text{Taylor}_a(f) = f(a) + f'(a)(x - a) + \frac{f''(a)}{2!}(x - a)^2 + \cdots .$$
>
> We can also write it as
>
> $$\text{Taylor}_a(f) = f^{(0)}(a) + f^{(1)}(a)(x - a) + \frac{f^{(2)}(a)}{2!}(x - a)^2 + \cdots$$
>
> or most compactly as
>
> $$\text{Taylor}_a(f) = \sum_{m=0}^{\infty} \frac{f^{(m)}(a)}{m!}(x - a)^m.$$

Taylor series are named after British mathematician Brook Taylor (1685–1731).

The definition of $\text{Taylor}_a(f)$ produces an infinite series, which is just a formal sum of terms. We should think of $\text{Taylor}_a(f)$ as a polynomial of infinite degree. We will try to sum $\text{Taylor}_a(f)$ and get a number when we substitute a value for x in the series, but we'll do that a little later.

Although the symbol f appears in the formal expression of the series, it is only there to tell you how to compute the coefficients of the series. The series itself has no f in it, just constants times powers of $(x - a)$. For example, there is the exponential function e^x, whose Taylor series at 0 is given by the infinite sum

$$1 + 1 \cdot (x - 0) + \frac{1}{2!} \cdot (x - 0)^2 + \frac{1}{3!} \cdot (x - 0)^3 + \cdots$$

or compactly as

$$\sum_{m=0}^{\infty} \frac{1}{m!} x^m.$$

We make an important convention: If a term in the Taylor series has coefficient 0, we don't need to write it. If all the terms from degree M onwards have coefficient 0, we don't write them, and the Taylor series looks like a polynomial of degree $M - 1$. In fact, it *is* a polynomial of degree $M - 1$.

EXERCISE: If $p(x)$ is a polynomial of degree d, then its Taylor series at a is again a polynomial of degree d.

EXERCISE: If $p(x)$ is a polynomial of degree d, then its Taylor series at 0 is equal to $p(x)$ itself.

WARNING: As a general rule, it is not possible to determine a function f by starting with Taylor$_a(f)$. The standard example of this is the function

$$f(x) = \begin{cases} e^{-1/x^2} & x \neq 0 \\ 0 & x = 0. \end{cases} \tag{12.2}$$

In a calculus class, one can prove that all the derivatives of $f(x)$ exist everywhere, even at $x = 0$, and compute that $f^{(m)}(0) = 0$ for every positive integer m. Therefore, the Taylor series of f at a is 0, that is, the series every term of which is 0. This certainly does not determine f, since the zero-function, whose value is 0 everywhere, also has its Taylor series at 0 equaling 0. In fact, there are infinitely many functions all of whose Taylor series at $a = 0$ are the 0-series.

However, this ambiguity does not occur with polynomials. If we know that $p(x)$ is a polynomial, and we are told Taylor$_a(p)$, then we know $p(x)$, because in fact $p(x)$ and Taylor$_a(p)$ define the same function of x.

You might wonder: Perhaps *all* the Taylor series of the function $f(x)$ in (12.2) at *all* possible a's together determine f. Well, they do, but this is sort of stupid. The constant term of $\text{Taylor}_a(f)$ is always $f(a)$, so knowing *all* the Taylor series of f is overkill as far as knowing or predicting f goes.

Let's summarize. First, let's define the concept of a power series, that is, a formal expression in powers of the variable, which looks like an infinitely long polynomial:

$$a_0 + a_1 y + a_2 y^2 + a_3 y^3 + \cdots .$$

Here, the variable is y, and the coefficients a_0, a_1, a_2, ... are constants. If a coefficient is 0, we don't bother to write the term, but just leave it out. There is an exception: If *all* the coefficients are 0, we have to write something, and so we write the zero power series as 0.

If a function $f(x)$ is infinitely differentiable at a, then the Taylor series at a of $f(x)$ is given by a formula in terms of the values of the derivatives of $f(x)$ at a. As we've said, that formula is

$$\text{Taylor}_a(f) = \sum_{m=0}^{\infty} \frac{f^{(m)}(a)}{m!}(x - a)^m .$$

Then $\text{Taylor}_a(f)$ is a power series in the variable $y = x - a$. If $f(x)$ is a polynomial, then from knowledge of $\text{Taylor}_a(f)$ for any single value of a, we can recover complete knowledge of f. This is action at distance: The derivatives of f at a, including the 0-th derivative (i.e., the value $f(a)$), completely determine the values of f everywhere, even very far away from a.

But that is when f is a polynomial. For a general function f, we have no such action at a distance. It should seem reasonable that if we knew of other functions, beyond polynomials, whose Taylor series also have this action-at-a-distance power, this would be valuable knowledge. We do, and it is. Such functions are called *analytic functions*.

3. Analytic Functions

Because this is not a treatise on calculus of real variables, or complex variables, we can't develop the entire fascinating theory of analytic functions. Instead, we cut to the chase, which means moving over from *real*-valued functions of a *real* variable to *complex*-valued functions of a *complex* variable. It is the field of complex numbers in which it is most natural to study analytic functions.

For the remainder of this chapter, our variables will be allowed to take complex values. The functions we talk about will also have complex values. We will also discuss polynomials that might have complex coefficients.

Don't forget that there is a difference between a polynomial and a function. A *polynomial* is a kind of symbolic expression, for example, $f(x, y, z) = 2x^3 - 3xy + z$. A *function* has a source and a target. If $f : U \to V$ is a function with source U and target V, then for any element u of the set $U, f(u)$ is a particular fixed element of V. Of course, a polynomial can always be interpreted as a function, because you can give values to the variables and get a value for the function. In our example, $f(1, 2, 3) = 2 \cdot 1^3 - 3 \cdot 1 \cdot 2 + 3 = -1$. In this case, $f : \mathbf{C}^3 \to \mathbf{C}$. (The domain $\mathbf{C}^3$ is the set of all ordered triples of complex numbers.)

When we look at a polynomial, the context should make it clear whether we are dealing with it as a symbolic expression or as a rule defining a function. Unfortunately (or fortunately), mathematical tradition allows us to use the same symbol, $f = f(x, y, z)$ in our example, for both possibilities.

Now we have to define a complex derivative. That's easy: We just use the same definition as in calculus, except that h can be any small *complex* number. Let $f(z)$ be a function $f : U \to \mathbf{C}$, where U is a small open disc in the complex plane centered at a. (An *open disc* centered at a of radius $r > 0$ is the set of all complex numbers z satisfying $|z - a| < r$.) Then we define

$$f'(a) = \lim_{h \to 0} \frac{f(a + h) - f(a)}{h}$$

if (and this is a big "if") that limit exists and equals the same complex number no matter how h tends to 0 in the complex plane.

The usual rules for differentiating polynomials also work for polynomials with complex coefficients. However, there are nonpolynomial functions that have derivatives when their domain is $\mathbf{R}$ but not when their domain is $\mathbf{C}$. For example, the function defined in equation (12.2) does not have a derivative at 0 if its domain is $\mathbf{C}$ rather than $\mathbf{R}$.

DEFINITION: We say that f is *analytic in U* if $f'(a)$ exists for every a in U. We say that f is *analytic at a* if it is analytic in some open disc U centered at a.

There is an enormous theory of analytic functions. It is so important that in the old days, the theory of analytic functions was simply called "function theory." These days, we are more egalitarian, and realize that there are lots of important functions that are not analytic.

Here are some examples of analytic functions. Directly from the definition, you can see that if c is any complex number, then the function $f(z) = c$ is analytic. (Its derivative is the 0-function.) Just as easily, you can see that $f(z) = z$ is analytic. (Its derivative is the constant function 1.) Imitating standard proofs from calculus, you can see that if f and g are analytic at a, so are $f + g, f - g$, and fg. If $g(a) \neq 0, f/g$ is also analytic at a. The chain rule also applies to analytic functions. It follows immediately from these facts that any polynomial is analytic everywhere.

DEFINITION: A function $f : \mathbf{C} \to \mathbf{C}$ that is analytic everywhere is called an *entire function*.

There are many entire functions other than polynomials. For example, e^z is entire. (By definition, if $z = x + iy$, $e^z = e^x(\cos y + i \sin y)$.) Because e^z is entire, you know that e^z times any polynomial is entire, and that e^z divided by a polynomial $p(z)$ is analytic at any a where $p(a) \neq 0$.

The main example of entire functions in this book, which we will see in chapter 13, are the L-functions of elliptic curves. They are always entire.

One of the main theorems about analytic functions says that if $f(z)$ is analytic in the open disc U, then $f'(z)$ is also analytic in U. Iterating this theorem, we see that *all* the (complex) derivatives of $f(z)$ exist in U. If U is centered at a, we can then write down the Taylor series of any analytic function $f(z)$. It is given by the same formula as before:

$$\text{Taylor}_a(f) = \sum_{m=0}^{\infty} \frac{f^{(m)}(a)}{m!}(z - a)^m.$$

Again, $\text{Taylor}_a(f)$ is a power series, but now the variable z is a complex number and a may be any complex number.

Calculus for dummies (see section 4 of chapter 4) also works for Taylor series. That is, if you want to find $\text{Taylor}_a(f')$, you simply use the rule that the derivative of $b(z-a)^m$ is $mb(z-a)^{m-1}$ and differentiate term by term.

EXERCISE: Prove that if

$$\text{Taylor}_a(f) = \sum a_n(z-a)^n,$$

then

$$\text{Taylor}_a(f') = \sum \frac{d}{dz}a_n(z-a)^n = \sum na_n(z-a)^{n-1}.$$

SOLUTION: By definition,

$$\text{Taylor}_a(f') = \sum_{m=0}^{\infty} \frac{(f')^{(m)}(a)}{m!}(z-a)^m.$$

But the m-th derivative of the first derivative is the $m+1$-st derivative. So

$$\text{Taylor}_a(f') = \sum_{m=0}^{\infty} \frac{f^{(m+1)}(a)}{m!}(z-a)^m.$$

On the other hand, if we differentiate $\text{Taylor}_a(f)$ term by term in "dummy" fashion, we get

$$\left[\sum_{m=0}^{\infty} \frac{f^{(m)}(a)}{m!}(z-a)^m\right]^{\text{diff. term by term}} = \sum_{m=1}^{\infty} \frac{f^{(m)}(a)}{m!}m(z-a)^{m-1}.$$

We can rewrite the last infinite sum by changing summation variables. We set $k = m - 1$, and use the fact that $m/m! = 1/k!$. The result is the infinite sum

$$\sum_{k=0}^{\infty} \frac{f^{(k+1)}(a)}{k!}(z-a)^k$$

and this last sum is just what we saw $\text{Taylor}_a(f')$ to be, with the summation variable called k rather than m.

By this time, you may be itching to sum the infinite series for each value of z, so that it will define a function of z. Of course, we cannot simply add up an infinite number of numbers. But we can agree that if the limit of $\sum_{m=0}^{n} a_m(z_0 - a)^n$ exists as $n \to \infty$, then we will say that the infinite series $\sum_{m=0}^{\infty} a_m(z - a)^n$ has a value at $z = z_0$, namely that limit. When this happens, we say that the power series *converges* at z_0.

For example, if we look at the Taylor series of e^z at 0, which is $\sum_{m=0}^{\infty} \frac{1}{m!} z^m$, and if we set $z = z_0$ for any complex number z_0, then the limit of the partial sums exist, and this limit happens to *equal e^{z_0}*. This is not a coincidence:

THEOREM 12.3: If $f(z)$ is analytic in an open disc U centered at a, then its Taylor series at a converges at z_0 for any z_0 in U, and the value of the infinite sum (i.e., the limit of the partial sums) equals $f(z_0)$.[1]

Theorem 12.3 implies action-at-a-certain-distance. If you know the Taylor series of the analytic function $f(z)$ at a, then you know for sure all the values of $f(z)$ in some open disc U around a. This disc U could be big, or even could be the whole complex plane (which we think of as an honorary disc).

For example, the Taylor series of e^z at 0 converges in the whole complex plane, and therefore information at one point (0) determines the values of e^z in the whole plane. This is truly an example of action at a distance. (The same is true for any entire function.)

Or the disc U could be rather small. For example, the function

$$f(z) = \frac{1}{z - 0.1}$$

---

[1] Something stronger is true, which we mention in case you know these terms. The Taylor series converges *absolutely* for any z_0 in U, and if V is any closed subdisc of U, the Taylor series converges *absolutely uniformly* on V. These extra properties of the Taylor series are important for proving things about analytic functions.

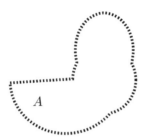

Figure 12.1. A typical open set

is analytic in the open disc of radius 0.1 centered at 0. The Taylor series at 0 of this function converges only in this disc, and not in any larger open disc centered at 0.

It could be very difficult to do these infinite summations. But the moral of the story is that if a function $f(z)$ is analytic on an open disc U centered at a, then its Taylor series at a determines f on U. The important thing to see here is that knowing the Taylor series is knowing just a list of complex numbers, namely the coefficients $a_0, a_1, a_2, \ldots$. This is just a countable set of data. But that knowledge determines the values of f at every point of U, which is an uncountable number of points. And U might be as large as the whole complex plane!

4. Analytic Continuation

Discs are not the be-all and end-all of shapes in the plane. By definition, an *open set* in the plane is a subset of the plane with the property that if it contains a point a then it also contains some open disc centered at a. This open disc might be very small, but it is there. In particular, the whole plane is an open set, and the empty set is also an open set. You could have an open set A like figure 12.1 and you could have a function $f : A \rightarrow \mathbf{C}$, that is analytic at every point of A. What can we say about the Taylor series?

If a is a point of A, then there will be a largest open disc U centered at a and contained in A, as shown in figure 12.2. By our theorem, the Taylor series of f at a will converge in U. Will it perhaps converge in a somewhat larger open disc centered at a? Maybe it will. If so, we can increase A, making it bigger by adding to it this larger open disc. We get a larger

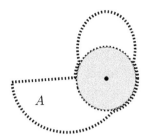

Figure 12.2. A region of convergence

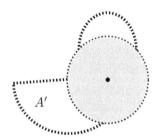

Figure 12.3. A larger region of convergence

domain A' for the function f, as in figure 12.3. We can keep doing this, by looking at another point a' in A', getting larger and larger domains A'', A''', etc., to which we can extend the function f, keeping it defined and analytic on the whole larger domain.

In this way, we can get a maximal domain to which f can be extended. There are some problems that arise that are beyond the scope of this book. There may not be a single natural maximal domain, because you might go around in a circle and when you get back to where you started from, the value of f may not equal what it was at the beginning. This problem is called *monodromy*. To deal with monodromy, you have to leave the world of the plane and start pasting discs together in some abstract world. Then you do get a natural maximal domain, called the *Riemann surface* of f. The complex points of an elliptic curve, described in section 6 of chapter 8, also give an example of a Riemann surface, as does the Riemann sphere.

In the cases of most importance to us, namely L-functions, we will begin not with a disc but with a half-plane. The maximal domain to which the L-function will extend will be the whole complex plane. The L-function, so extended, will be an entire function.

The process of extending f to its maximal domain is called "analytic continuation." You can think of the process like this. You plant a seed somewhere. This seed is the Taylor series of a function f that is analytic at a, and we plant it at a. So f starts off as an analytic function in some open set (perhaps very small) containing a. You may not even be given the Taylor series of f; the function might be defined in some other way. But because f is analytic at a, you know that f has a Taylor series at a and is *determined* by its Taylor series at a. The theory of analytic continuation tells you that f can be extended to a function on some maximal domain M. We generally use the same letter for the extended function. So now $f : M \to \mathbf{C}$ is the analytic function extended to this possibly large domain.

The main point: *The values of f everywhere in M are determined by the Taylor series* $\mathrm{Taylor}_f(a)$ *with which we started.* This is because at each step of the analytic continuation process, we extend f by adding the points of a new disc to the domain of f, and the values of f at the new points are determined by the Taylor series of f at the center of the new disc, and the Taylor series of f at the center of the new disc is determined by the values of f and its derivatives at that center, and they in turn are determined by the values of f in some teeny disc around that center, and those values are determined by the f we already knew in the previous step of the process. (Take a deep breath and reread the previous two sentences.)

Working backwards, the values of f in the new disc are determined by the values of f in some previous disc, and eventually we get back to the values of f in the first little domain we started with. Another metaphor for analytic continuation would be the growth of a crystal.

This is action at a distance with a vengeance. We can begin with a very small open disc—around 0 say—and an analytic function $f(z)$ defined by hook or by crook in that disc. Suppose that $f(z)$ analytically continues to become an entire function in the whole plane. First of all, we may not have guessed that $f(z)$ would determine any kind of function far away from $z = 0$. But it does. Calling this extended function $f(z)$ again, we then know that the value of $f(z_0)$ is determined by the embryonic f we started with. For example, $f(100 + 200i)$ is determined by the values of f in that initial disc of very small radius around 0.

But—and this is a big "but"—it can be very tricky to get a handle on what number $f(100 + 200i)$ actually is, or even whether it is 0 or not.

Here is a good illustration of all this: Let's review the definition of the Riemann zeta-function from p. 169. Recall that $n^s = e^{s \log n}$. Because of the chain rule, n^s is an entire function, analytic for all values of s. We form the sum:

$$Z_k(s) = \frac{1}{1^s} + \frac{1}{2^s} + \frac{1}{3^s} + \cdots + \frac{1}{k^s}.$$

(Of course $\frac{1}{1^s}$ is just the constant function 1, but we wrote it that way to enforce the pattern.)

Since $Z_k(s)$ is a finite sum of analytic functions, it is analytic. Now it is not always true that an *infinite* sum of analytic functions is analytic. For one thing, if the infinite sum doesn't converge, it's not clear how it would define a function at all. However, it happens that if you take any open set A on which the real part of s stays greater than 1, no matter what the imaginary part of s is doing, then the limit of $Z_k(s)$ as $k \to \infty$ exists for each s in A and defines an analytic function on A.[2] This limit function is $\zeta(s)$, the Riemann zeta-function:

$$\zeta(s) = \frac{1}{1^s} + \frac{1}{2^s} + \frac{1}{3^s} + \cdots.$$

OK, we haven't done any analytic continuation yet. It turns out (as we claimed on p. 173) that we can, and that the maximal domain of $\zeta(s)$ is $A = \mathbf{C} - \{1\}$, the whole complex plane with the number 1 removed. The Riemann Hypothesis says that any value $s = x + iy$ for which $\zeta(x + iy) = 0$ (we're talking about the extended function here) satisfies *either* $y = 0$ and x is a negative even integer *or* $x = 1/2$. (The standard reference is (Titchmarsh, 1986), and a good popular history is (Derbyshire, 2004).) This is another one of those problems for whose solution the Clay Institute will pay a million dollars.

What makes the Riemann Hypothesis so hard to prove or disprove is the mysterious nature of analytic continuation. The function $\zeta(s)$ is completely determined by its values in a small disc about $s = 2$ (for example) but *how* exactly this determination goes is opaque. We will see the same type of phenomenon when we look at L-functions of elliptic curves.

---

[2] For those who knew the terms in footnote 1, we can say that if a series of analytic functions all defined on the same open set A converges absolutely uniformly in a neighborhood of each point of A, then the limit function is analytic on A.

5. Zeroes, Poles, and the Leading Coefficient

The Riemann zeta-function has lots of properties. Here's one: If you multiply $\zeta(s)$ by $s - 1$, then you get a new function, $(s - 1)\zeta(s)$. This new function has its analytic continuation *everywhere* to the whole s plane. In other words, it is an entire function.

This means something nice: We couldn't talk about the Taylor series of $\zeta(s)$ at $s = 1$ because ζ wasn't even defined at $s = 1$. But we can discuss the Taylor series of $(s - 1)\zeta(s)$ at $s = 1$. Let $f(s) = (s - 1)\zeta(s)$, and then

$$\text{Taylor}_1(f) = a_0 + a_1(s - 1) + a_2(s - 1)^2 + a_3(s - 1)^3 + \cdots \quad (12.4)$$

for some complex numbers $a_0, a_1, \dots$. (It turns out that $a_0 = 1$ and the other coefficients can be expressed using complicated formulas that are not worth writing down here.)

Equation (12.4) is very nice even if we don't know the numbers $a_0, a_1, \dots$. We can write some kind of series for the Riemann zeta-function itself, by dividing both sides by $(s - 1)$:

$$\zeta(s) = \frac{a_0}{s - 1} + a_1 + a_2(s - 1) + a_3(s - 1)^2 + \cdots.$$

The right-hand side is like a power series in powers of $s - 1$, except that some (in this case one) of the powers are negative. Series like that are called *Laurent series*. They were first described in print by French mathematician Pierre Alphonse Laurent (1813–54).

Suppose $f(z)$ is a nonzero function which is analytic at a with Taylor series

$$\text{Taylor}_a(f) = a_0 + a_1(z - a) + a_2(z - a)^2 + a_3(z - a)^3 + \cdots.$$

Because f is not the zero function, and because the Taylor series determines f, at least one coefficient a_k must be nonzero. Let k be the smallest integer for which $a_k \neq 0$. In other words, we can write

$$\text{Taylor}_a(f) = a_k(z - a)^k + a_{k+1}(z - a)^{k+1} + a_{k+2}(z - a)^{k+2} + \cdots.$$

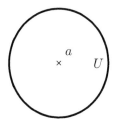

Figure 12.4. A small disk U missing its center

DEFINITION: In these circumstances, we say $f(z)$ has a *zero of order k* at a. Its *leading term* at a is $a_k(z - a)^k$. Its *leading coefficient* at a is a_k.

Our definition leads to the peculiar usage that if f is not zero at all at a, that is, if $f(a) \neq 0$, we can still say that $f(z)$ *has a zero of order* 0 at a. This usage allows us to state general theorems more easily.

Now suppose that $f(z)$ is a nonzero function, not analytic at a, but analytic near a, all around a. In other words, let U be a small disc centered at a and suppose A is that disc, but with the center a removed, as in figure 12.4. In other words, $A = U - \{a\}$. And suppose that $f(z)$ is analytic on A, but cannot be extended to be analytic on all of U. We say f has a *nonremovable singularity* at a. This is the case with $\zeta(s)$ for $s = 1$.

We'd still like to have something like a Taylor series at a for $f(z)$ if possible. The general approach is more complicated than what we need. For our purposes, we will assume that there is some positive integer k with the property that $(z - a)^k f(z)$ can be extended to be analytic on the whole disc U. (For example, for the zeta-function, we took $k = 1$ and $a = 1$.) We will also assume that k is the smallest such integer for which $(z - a)^k f(z)$ is analytic. Then we can write:

$$\text{Taylor}_a((z - a)^k f(z)) = a_0 + a_1(z - a) + a_2(z - a)^2 + a_3(z - a)^3 + \cdots .$$

with $a_0 \neq 0$.

We can now divide both sides by $(z - a)^k$ and get a sort of power series, with some negative exponents, that converges to the function $f(z)$ on A. In other words, we can write

$$f(z) = a_0(z - a)^{-k} + a_1(z - a)^{-k+1} + a_2(z - a)^{-k+2} + \cdots .$$

DEFINITION: In these circumstances, we say $f(z)$ has a *pole* of order k at a. Its *leading term* at a is $a_0(z - a)^{-k}$. Its *leading coefficient* at a is a_0. The series on the right-hand side of the equation is called the *Laurent series* of $f(z)$ around $z = a$.

EXAMPLE: The Riemann zeta-function $\zeta(s)$ has a pole of order 1 at $s = 1$ with leading coefficient 1.

When we analytically continue functions, the continuation might have poles. However, a pole is not too scary, because a pole at a can be dealt with by multiplying the function by a power of $z - a$ and using Taylor series. The L-functions we'll be looking at later actually analytically continue without developing poles—they analytically continue to be entire functions.

Chapter 13

L-FUNCTIONS

Road Map

In this chapter, we see how to use elliptic curves to build a type of generating function called an *L*-function. The hope is that important properties of the elliptic curve will be captured by computable properties of the *L*-function.

1. A Fertile Idea

The concept and notation for an *L*-function seem to go back to Lejeune Dirichlet, in his famous paper from 1837, "Beweis eines Satzes über die arithmetische Progression." In this paper, he proved that if you start with two positive integers a and b that share no common prime factor, then the sequence

$$a, a + b, a + 2b, a + 3b, a + 4b, \ldots$$

contains infinitely many prime numbers. We won't describe the proof here; it uses a function that Dirichlet denoted by L and which was later called $L(\chi, s)$. We will describe this function and some of its more general cousins at the end of this chapter.

There are many mansions in the house of *L*-functions. There are *L*-functions associated to linear Galois representations, like Dirichlet's *L*-functions. There are *L*-functions associated to modular forms and automorphic forms. There are *L*-functions associated to projective curves and more generally to algebraic varieties defined over **Z**. And there are various generalizations of all these *L*-functions. Many of these objects are explained in *Fearless Symmetry*.

All these different kinds of L-functions are related in various ways. Some of these relations are proven theorems, and some of them are basic conjectures in modern number theory that drive much of the current research in that field. When a relationship between different kinds of L-functions holds, it provides a link of great interest in itself, which is also useful for proving further theorems. For example, the fact that the L-function of an elliptic curve is equal to the L-function of a certain modular form is the essence of the great theorem proved by the cumulative work in (Wiles, 1995; Taylor and Wiles, 1995; Breuil et al., 2001). In turn, that equality can be used to prove Fermat's Last Theorem, as British mathematician Andrew Wiles (1953–) did.

Dirichlet's L-functions can be thought of as a generalization of the Riemann zeta-function $\zeta(s)$. In the next section, we will describe a monster generalization of $\zeta(s)$ called the Hasse–Weil zeta-function. (The German mathematician Helmut Hasse lived from 1898 to 1979. The French mathematician André Weil lived from 1906 to 1998.) This thing generally breaks up into a product of L-functions (and zeta-functions) and their reciprocals, so it can fruitfully be viewed as one source of L-functions. This is the kind of L-function that will be most important to us in the formulation of the Birch–Swinnerton-Dyer Conjecture. We will only describe the Hasse–Weil zeta-function for a nonsingular projective plane curve defined by a homogeneous equation $F(x, y, z) = 0$ with coefficients in $\mathbf{Q}$. That will be enough for now.

2. The Hasse–Weil Zeta-Function

Suppose we have a homogeneous equation $F(x, y, z) = 0$ with coefficients in $\mathbf{Q}$. Suppose the degree of F is $d \geq 1$ and that the projective curve C defined by this equation is nonsingular. Multiplying through by a common denominator, we may assume that the coefficients of F are actually in $\mathbf{Z}$. We can then reduce them modulo p and obtain a polynomial $\overline{F}(x, y, z)$ with coefficients in the finite field $\mathbf{F}_p$ that has p elements. (We view $\mathbf{F}_p$ and $\mathbf{Z}/p\mathbf{Z}$ as the same object.)

Solutions to $\overline{F}(x, y, z) = 0$, where x, y, and z are set equal to elements of $\mathbf{F}_p$, are the same as solutions to the congruential equation $F(x, y, z) \equiv 0 \pmod{p}$ after we reduce the coordinates modulo p. In more detail: For any integer u, let's write $\bar{u}$ for the element in $\mathbf{F}_p$ which is the remainder we get when dividing u by p. If a, b, and c are integers and $F(a, b, c) \equiv 0 \pmod{p}$, then $\bar{F}(\bar{a}, \bar{b}, \bar{c}) = 0$ in $\mathbf{F}_p$.

> **EXAMPLE:** Let $F(x, y, z) = y^2 z - x^3 + xz^2$. Let $x = 11$, $y = 12$, and $z = 1$. Then $F(11, 12, 1) = 12^2 - 11^3 + 11 = 144 - 1331 + 11 = -1176$. So $(3, 2, 1)$ is not a solution to $F = 0$. However, $-1176 \equiv 0 \pmod{7}$. So if we set $p = 7$, then $F(11, 12, 1) \equiv 0 \pmod{7}$. Therefore $(\overline{11}, \overline{12}, \overline{1}) = (\overline{4}, \overline{5}, \overline{1})$ is a solution to $\bar{F}(x, y, z) = y^2 z - x^3 + xz^2 = 0$ in $\mathbf{F}_7$.
>
> In more geometric terms, note that the projective curve defined by $y^2 z - x^3 + xz^2 = 0$ is an elliptic curve E with $\Delta_E = 64$. So we can say that $(\bar{4} : \bar{5} : \bar{1})$ is an element of the projective curve $E(\mathbf{F}_7)$ defined over the field $\mathbf{F}_7$, or, more briefly, that it is a point on the elliptic curve $E(\mathbf{F}_7)$.
>
> We can do more with this example, because -1176 is also congruent to 0 modulo 3. Setting $p = 3$ now, we can say that $(\overline{11} : \overline{12} : \overline{1}) = (\bar{2} : \bar{0} : \bar{1})$ is a point on the elliptic curve $E(\mathbf{F}_3)$. Same E, different p. Note that the overline notation $\bar{u}$ for an integer u depends crucially on which prime p we are "modding out" by, which must be picked up from the context.

> **EXERCISE:** Make up an elliptic curve over $\mathbf{Q}$ and find some points on it modulo different primes.

Our example and exercise involved elliptic curves because they are the main characters of this book. But we will continue to explain the zeta-function more generally for any nonsingular plane projective curve C, given by a homogeneous equation $F(x, y, z) = 0$ with integer coefficients, of degree d. Now the nice thing about reducing F modulo p is that in the equation $\bar{F}(x, y, z) = 0$, we can plug in values for x, y, and z from *any* field containing $\mathbf{F}_p$ (i.e., any field of characteristic p). In particular, we know there is a finite field with p^n elements for any positive integer n, which we designated by the symbol $\mathbf{F}_{p^n}$. (See section 7 of chapter 2.) So we can look at solutions (a, b, c) to $\bar{F}(x, y, z) = 0$ where a, b, and c are all in some $\mathbf{F}_{p^n}$.

Let's temporarily fix a p and look at the sequence of finite fields $\mathbf{F}_p, \mathbf{F}_{p^2}, \mathbf{F}_{p^3}, \ldots$. Correspondingly, we can find all the points on

$C(\mathbf{F}_p)$, $C(\mathbf{F}_{p^2})$, $C(\mathbf{F}_{p^3})$, ... For a single $C(\mathbf{F}_{p^n})$, this is a finite-length calculation that is not too hard to program a computer to do, once the computer
knows how to do arithmetic in the finite field $\mathbf{F}_{p^n}$. The arithmetic can be
specified by an addition table and a multiplication table worked out once
and for all, and stored. Let's count these points. Say there are M_1 points on
$C(\mathbf{F}_p)$, M_2 points on $C(\mathbf{F}_{p^2})$, M_3 points on $C(\mathbf{F}_{p^3})$, and so on. (Don't forget
to count the points at infinity.)

Remember the discussion of generating functions in chapter 11? This
sequence of nonnegative integers M_1, M_2, M_3, ... is a perfect grist for
some generating function mill. Which one? Experience shows that we
should package this data using the ideas of section 6 of chapter 11.

To review, that means that we construct

$$\zeta_{C,p}(T) = \exp\left(\sum_{r=1}^{\infty} \frac{M_r T^r}{r}\right)$$

There is an amazing theorem that says that there exists two *polynomials*
$r_{C,p}(T)$ and $q_{C,p}(T)$ that have integer coefficients with the property that

$$\zeta_{C,p}(T) = \frac{r_{C,p}(T)}{q_{C,p}(T)}.$$

Now a polynomial only has a finite number of coefficients. So this theorem
tells us that the *infinite* sequence M_1, M_2, M_3, ..., can be computed from
the finite amount of data in the polynomials $r_{C,p}(T)$ and $q_{C,p}(T)$ by using
theorem 11.20. To review:

Step 1: Factor $r_{C,p}(T) = (1 - \beta_1 T)(1 - \beta_2 T) \cdots (1 - \beta_j T)$.
Step 2: Factor $q_{C,p}(T) = (1 - \alpha_1 T)(1 - \alpha_2 T) \cdots (1 - \alpha_k T)$.
Step 3: Then $M_r = \alpha_1^r + \alpha_2^r + \cdots + \alpha_k^r - \beta_1^r - \beta_2^r - \cdots - \beta_j^r$.

It follows from this recipe that given the curve C, in theory we only need to
work out how many points are on $C(\mathbf{F}_{p^n})$ for a finite number of the fields
$\mathbf{F}_{p^n}$ in order to be able to predict how many points there are on all the other
$C(\mathbf{F}_{p^n})$'s!

The theory gives even more information. The curve C has a number
attached to it, called the genus, g, which can be computed from the
degree. The genus has a nice topological characterization: If you look at

the complex points $C(\mathbf{C})$, then as a topological space $C(\mathbf{C})$ is the surface of a "doughnut" with g holes. (For example, a torus has genus 1.) The higher the degree d, the higher the genus g. A projective line has $g = 0$, and an elliptic curve has $g = 1$. Now the theorem mentioned above specifies further that $r_{C,p}(T)$ has degree $2g$ and $q_{C,p}(T) = (1 - T)(1 - pT)$. Thus for a projective line, we need no information to determine the M_n's (see the next example.) And for an elliptic curve, the value of M_1 determines all the M_n's. This will be apparent again later in this chapter, when we define the L-function of an elliptic curve, which will depend only on the M_1's for different p's.

> **EXAMPLE:** There is only one example we can do with our bare hands, because there is only one kind of nonsingular projective curve for which we know the M_n's without any work. Namely, let $C = P^1$ be the projective line, given by the equation $x = 0$. Here, $F(x, y, z) = x$ is homogeneous of degree 1 and P^1 has genus $g = 0$. Therefore, by the theorem we've been talking about, for each p we should have $\zeta_{P^1,p}(T) = 1/(1 - T)(1 - pT)$. Let's check this.
>
> How many points are on the projective line $P^1(K)$ for any finite field K? The projective line consists of the usual y-axis plus one point at infinity. If K has k elements, then the usual line has just one coordinate that can take any of these k values. We add 1 for the point at infinity. So the number of points on $P^1(K)$ is $k + 1$.
>
> Now fix p and note that by definition, $\mathbf{F}_{p^r}$ has p^r elements. So $M_r = p^r + 1$ for all r. (This verifies the prediction that we don't need to know anything special to figure out the M_r's. They are all predictable *a priori*. We don't even have to know M_1. It too is predictable, unlike the case of elliptic curves.)
>
> Alternatively, we can apply Theorem 11.20. If $f(T) = 1$, and $g(T) = (1 - T)(1 - pT)$, then we know that there are no β's to worry about, $\alpha_1 = 1$, and $\alpha_2 = p$. The theorem lets us conclude that

$$
\exp\left(\sum_{r=1}^{\infty} \frac{(p^r + 1)T^r}{r} \right) = \frac{1}{(1 - T)(1 - pT)} = \zeta_{P^1,p}(T),
$$

and therefore $M_r = 1^r + p^r$. This agrees with what we found by direct calculation.

For each p, we call the function $\zeta_{C,p}(T)$ the *local zeta-function* of C. Now we can define the Hasse–Weil zeta-function of C, denoted $Z(C, s)$, except for a certain fudge factor we will explain but not define. You simply substitute $T = p^{-s}$ in $\zeta_{C,p}(T)$ and then multiply all these local factors together:

$$Z(C, s) = \prod_{p \text{ good}} \zeta_{C,p}(p^{-s}) \times \text{fudge}.$$

Here is a rough idea of where the fudge factor comes from. We call a prime p "good" for C if C modulo p is nonsingular, which means $C(\mathbf{F}_p^{\text{ac}})$ has no singular point on it, where $\mathbf{F}_p^{\text{ac}}$ is as usual an algebraic closure of $\mathbf{F}_p$. We call p "bad" if it is not good. For a given curve C, there are only a finite number of bad primes. The factor called "fudge" is a product of certain functions of p^{-s} for the bad primes p. These functions are defined in a complicated way we can't go into here. (However, when C is an elliptic curve, we will be able to tell you how to define the fudge factors.) The fudge factors are hard to define, but they are essential to flesh out our understanding of C modulo p for *all* primes p. In particular, they are necessary to make a certain functional equation discussed below work out correctly.

Let's continue our example with $C = P^1$, a projective line. To get the Hasse–Weil zeta-function of P^1, we have to substitute $T = p^{-s}$ and then multiply all these local factors together:

$$Z(P^1, s) = \prod_p (1 - p^{-s})^{-1}(1 - pp^{-s})^{-1} \tag{13.1}$$

There are no bad primes in this example, because P^1 modulo p is nonsingular for all p. If we compare equation (13.1) with the Euler product of the Riemann zeta-function in equation (11.13), we see that

$$Z(P^1, s) = \zeta(s)\zeta(s - 1).$$

(Note that $pp^{-s} = p^{1-s} = p^{-(s-1)}$.)

EXERCISE: The Hasse–Weil zeta-function can be defined for any system of polynomial equations with integer coefficients (called a Z-variety). Write down the definition and compute the Hasse–Weil zeta-function for the 0-dimensional variety consisting of a single point.

SOLUTION: Given the **Z**-variety V, for every prime p, set M_n to be the number of points on $V(\mathbf{F}_{p^n})$ and define $\zeta_{(V,p)}$ exactly as we did for curves. Then define $Z(V, s)$ exactly the same way as we did for curves. We are not telling you how to figure out for which p the variety V (mod p) is nonsingular, nor are we telling you how to compute the Euler factors for the bad p's, but we won't need to know those things to do the exercise.

A single point would be given by the equation $f(x, y) = x = 0$ for $(x : y)$ in the projective line. Notice f is homogeneous of degree 1, and the coefficient 1 of x lies in **Z**. Let's call this variety V. Fix p. Then all the $M_n = 1$ because there is only a single point, and V (mod p) is nonsingular, whatever p may be. So we form

$$\zeta_{V,p}(T) = \exp\left(\sum_{r=1}^{\infty} \frac{T^r}{r}\right),$$

and we saw in equation (11.16) that $\zeta_{V,p}(T) = \frac{1}{1-T}$. To get the Hasse–Weil zeta-function, we have to substitute $T = p^{-s}$ and then multiply all these local factors together:

$$Z(V, s) = \prod_p (1 - p^{-s})^{-1}$$

which we recognize as the Euler product for the Riemann zeta-function from equation (11.13). So $Z(V, s) = \zeta(s)$.

Up to now, all of our manipulations have been formal algebraic matters, without worrying about whether the infinite series converge. For any **Z**-variety V, there is a real number σ such that the Euler product for the Hasse–Weil zeta-function of V converges absolutely for all s with $\Re(s) > \sigma$. There are conjectures about the analytic properties of the Hasse–Weil zeta-functions as you try to extend them to all values of s.

3. The *L*-Function of a Curve

Despite our simple examples above, the Hasse–Weil zeta-function of a **Z**-variety is usually not just a Riemann zeta-function or a product of several

Riemann zeta-functions. A very deep theorem gives more information about the local zeta-function of a general $\mathbf{Z}$-variety: It can be written as a quotient of two functions of p^{-s}, and the numerator and denominator of this quotient can be factored into a product of functions of p^{-s}. Each of these factors is the reciprocal of the L-function of a Galois representation, which we will define briefly in the last section of this chapter, for readers who know a little linear algebra.

We have already seen this factorization in the case of a nonsingular projective curve C of genus g. The numerator of the Hasse–Weil zeta-function of C is the product over good primes of polynomials of degree $2g$ evaluated at p^{-s}, times a fudge factor from the bad primes. The reciprocal of this numerator is called the L-function of C, and is written $L(C, s)$.

In the case of a curve, we have the formula

$$L(C, s) = \text{fudge} \times \prod_{p \text{ good}} \frac{1}{f_p(p^{-s})},$$

where $f_p(T)$ is some polynomial of degree $2g$. The theory tells us that the constant term of f_p is always equal to 1, and various theorems give us bounds on the absolute values of the coefficients of f_p.

So we can write this as

$$L(C, s) = \text{fudge} \times \prod_{p \text{ good}} \frac{1}{1 + c_{1,p}p^{-s} + c_{2,p}p^{-2s} + \cdots + c_{2g,p}p^{-2gs}}.$$

(Remember that although we don't say here what the fudge factor is, it is known and can be determined.) Using long division for each term in the product, divide the denominator into 1, and you will get a series in powers of p^{-s}, just as you do when dealing with the Riemann zeta-function. When you multiply all these series together for the different primes, you get a Dirichlet series, again just like the Riemann zeta-function:

$$L(C, s) = \sum_{n \geq 1} \frac{a_n}{n^s}$$

for some numbers a_n that depend on C. The bounds on the absolute values of the coefficients of f_p can be used to prove that this Dirichlet series converges in some right half-plane, as all Dirichlet series with number theoretic significance ought to do.

4. The *L*-Function of an Elliptic Curve

In this section, we will carry out the details of constructing the *L*-function of an elliptic curve *E* defined over **Q**. In this case, we can even say what the fudge factors are. In this section, and for the remainder of this book, we use the so-called "minimal model" for *E*. This is an equation for *E* with smallest possible $|\Delta_E|$ chosen among all possible equations defining *E*. Using the minimal model allows us to specify the fudge factors correctly.

Recall from chapter 8 that *E* (mod *p*) is an elliptic curve defined over **F**$_p$ for any *p* not dividing the discriminant Δ_E. That is, *E* modulo *p* is still a nonsingular projective curve, with $E(\mathbf{F}_p)$ nonempty, because the point at infinity on *E* will always reduce to a point modulo *p*. In fact, $E(\mathbf{F}_p)$ is an abelian group, which we have explained in detail. We have called those *p* that do not divide Δ_E the good primes of *E*, and those *p* that divide Δ_E the bad primes.

We use the symbol $E(\mathbf{F}_p)_{\text{ns}}$ for the set of nonsingular points on $E(\mathbf{F}_p)$. So if *p* is a good prime, $E(\mathbf{F}_p)_{\text{ns}} = E(\mathbf{F}_p)$. If *p* is a bad prime, then $E(\mathbf{F}_p)_{\text{ns}}$ has one fewer point than $E(\mathbf{F}_p)$, for we deduct the singular point (there is only one singular point). The set $E(\mathbf{F}_p)_{\text{ns}}$ is still nonempty, and is in fact also an abelian group, just as $E(\mathbf{F}_p)$ was for good *p*, as we saw in chapter 9.

We let N_p stand for the number of points in $E(\mathbf{F}_p)_{\text{ns}}$. For the good primes *p*, *E* modulo *p* is nonsingular, so N_p is simply the number of points in $E(\mathbf{F}_p)$. For the bad primes *p*, we worked out N_p in chapter 9. The results, taken from table 9.4, are that $N_p = p$ if *E* has additive reduction at *p*, $N_p = p - 1$ if *E* has split multiplicative reduction at *p*, and $N_p = p + 1$ if *E* has nonsplit multiplicative reduction at *p*.

Next we review the definition of the integer a_p, defined for every prime *p* . If *E* modulo *p* is nonsingular (good *p*), we set

$$N_p = p + 1 - a_p.$$

If E modulo p is singular (bad p), we set

$$N_p = p - a_p,$$

where we took away 1 because $E(\mathbf{F}_p)$ has one singular point that we are not counting.

For good p, the integer a_p is somewhat mysterious. Hasse proved (theorem 8.5) that $|a_p| \leq 2\sqrt{p}$ for every good p. The distribution of the a_p's within that range as p varies is addressed by the Sato–Tate Conjecture, formulated independently around 1960 by Japanese mathematician Mikio Sato (1928–) and Tate. This conjecture was proved recently by British mathematician Richard Taylor (1962–), building on the work of many mathematicians. But if you just pick a random p and ask what is a_p, it is hard to say without computing the number of points on $E(\mathbf{F}_p)$ (or without using the modular form attached to E, if you happen to know what that is).

For bad p, we know a_p exactly from table 9.4. Namely, $a_p = 0$ if E has additive reduction at p, $a_p = 1$ if E has split multiplicative reduction at p, and $a_p = -1$ if E has nonsplit multiplicative reduction at p.

We can now state a theorem first proved by Hasse. We have defined the L-function of a nonsingular projective plane curve above. If we apply this to E, we expect that the local zeta-function of E at a good p should depend on only two numbers, since E has genus 1. It turns out that we know *a priori* what the product of those two numbers is, so the local zeta-function of E at a good p should depend on only one number. In fact, it depends only on a_p and Hasse showed the formula that gives you the local zeta-function in terms of a_p. We also can determine the fudge factors for the bad primes, and putting it all together we obtain:

THEOREM 13.2: Let S be the set of primes for which $E(\mathbf{F}_p)$ is singular. Then

$$L(E, s) = \prod_{p \in S} \frac{1}{1 - a_p p^{-s}} \prod_{p \notin S} \frac{1}{1 - a_p p^{-s} + p^{1-2s}}.$$

It follows from the general theory that the value of a_p determines not only N_p but also the number of points on $E(\mathbf{F}_{p^n})$ for all n. Thus $L(E, s)$

encapsulates all the information about how many solutions there are to the cubic equation defining E over all finite fields. The BSD Conjecture is going to suggest that $L(E, s)$ also contains information about how many *rational* solutions there are to the cubic equation defining E. We are very close to the goal of explaining this.

Remember that we can do long division and multiplication and rewrite the product for $L(E, s)$ as a Dirichlet series:

$$L(E, s) = \sum_{n \geq 1} \frac{a_n}{n^s} = \sum_{n \geq 1} a_n n^{-s}.$$

The a_n's are all determined, of course, by the a_p's.

To see this explicitly, we will take a term in the infinite product and work out the first few terms of the long division. We have the formula for the infinite geometric series:

$$\frac{1}{1 - x} = 1 + x + x^2 + x^3 + \cdots .$$

This formula always holds in a formal sense, even if the right-hand side doesn't converge. ("Holds in a formal sense" means that if you start multiplying out $(1 - x)(1 + x + x^2 + x^3 + \cdots)$ and go on forever, you will just get 1.) So we can apply the geometric series equation for any x. If $x = a_p p^{-s}$, we get

$$\frac{1}{1 - a_p p^{-s}} = 1 + a_p p^{-s} + a_p^2 p^{-2s} + a_p^3 p^{-3s} + \cdots . \qquad (13.3)$$

If $x = a_p p^{-s} - p^{1-2s}$, we get

$$\frac{1}{1 - a_p p^{-s} + p^{1-2s}} = \frac{1}{1 - (a_p p^{-s} - p^{1-2s})}$$

$$= 1 + (a_p p^{-s} - p^{1-2s}) + (a_p p^{-s} - p^{1-2s})^2$$

$$+ (a_p p^{-s} - p^{1-2s})^3 + \cdots . \qquad (13.4)$$

We use the binomial theorem to expand $(a_p p^{-s} - p^{1-2s})^j$, and then rewrite $p^{m-ks} = p^m(p^{-ks})$. We get

$$\frac{1}{1 - (a_p p^{-s} - p^{1-2s})} = 1 + a_p p^{-s} - pp^{-2s} + a_p^2 p^{-2s} - 2a_p pp^{-3s}$$

$$+ p^2 p^{-4s} + a_p^3 p^{-3s} + \cdots$$

$$= 1 + \frac{a_p}{p^s} + \frac{a_p^2 - p}{p^{2s}} + \frac{a_p^3 - 2pa_p}{p^{3s}} + \cdots \qquad (13.5)$$

The definition of a_{p^2} is that it is the coefficient of p^{-2s}, so we can see from equation (13.5) that $a_{p^2} = a_p^2 - p$, and similarly that $a_{p^3} = a_p^3 - 2pa_p$.

We can start multiplying out these infinite series for various p to get terms in the Dirichlet series. For example,

$$(1 + a_3 3^{-s} - 3(3^2)^{-s} + a_3^2(3^2)^{-s} - \cdots)(1 + a_5 5^{-s} - 5(5^2)^{-s}$$

$$+ a_5^2(5^2)^{-s} - \cdots)$$

will give you some terms for your Dirichlet series:

$$1 + a_3 3^{-s} + (a_3^2 - 3)9^{-s} + a_5 5^{-s} + (a_5^2 - 5)25^{-s} + a_3 a_5 15^{-s} + \cdots .$$

Note that $a_{15} = a_3 a_5$.

> **EXERCISE:** Show that there is no ambiguity in our notation. That is, if $n = p$ is prime, then the a_n appearing in the infinite sum is the same number as the a_p appearing in the infinite product.

> **EXERCISE:** There is yet more potential ambiguity in our notation. We have defined a_p twice. The first time, we set $a_p = p + 1 - N_p$, and the second time, we have told you that a_p shows up in the formula for the numerator of $\zeta_{E,p}(T)$. Show that these two usages do not contradict each other.

> **SOLUTION:** Suppose that on the one hand, we define $a_p = p + 1 - N_p$, so that $N_p = p + 1 - a_p$. On the other hand,

suppose that we know that

$$\zeta_{E,p}(T) = \exp\left(\sum_{r=1}^{\infty} \frac{N_{p^r} T^r}{r}\right) = \frac{1 - a_p T + p T^2}{(1 - T)(1 - pT)}. \qquad (13.6)$$

In this formula, remember that $N_{p^r} = \#E(\mathbf{F}_{p^r})$.

Suppose that we factor the numerator of equation (13.6), and get

$$(1 - a_p T + p T^2) = (1 - \beta_1 T)(1 - \beta_2 T).$$

We know by multiplying these factors together that $\beta_1 \beta_2 = p$ and $\beta_1 + \beta_2 = a_p$. Theorem 11.20 now tells us that

$$N_{p^r} = 1^r + p^r - \beta_1^r - \beta_2^r.$$

In particular, $N_p = 1 + p - \beta_1 - \beta_2 = p + 1 - (\beta_1 + \beta_2) = p + 1 - a_p$.

EXERCISE: Verify the numerical claims in section 8 of chapter 8 that for the curve $y^2 = x^3 - x$, $a_9 = -3$, and for the curve given by $y^2 - y = x^3 - x$, $a_4 = 2$.

SOLUTION: We showed above that in general $a_{p^2} = a_p^2 - p$. The curve $y^2 = x^3 - x$ had $a_3 = 0$. The equation $a_9 = -3$ comes from the formula $a_{p^2} = a_p^2 - p$. For the curve $y^2 - y = x^3 - x$, $a_2 = -2$. The equation $a_4 = 2$ comes from the same formula.

The more interesting results in that section were counting the number of points on various curves. Return to $y^2 = x^3 - x$. We computed that $N_3 = 4$ and $a_3 = 0$. We now know more than we did earlier, so we can write

$$\zeta_{E,3}(T) = \exp\left(\sum_{r=1}^{\infty} \frac{N_{p^r} T^r}{r}\right) = \frac{1 + 3T^2}{(1 - T)(1 - 3T)}.$$

Factor the numerator: $1 + 3T^2 = (1 - \sqrt{-3}T)(1 + \sqrt{-3}T)$. Set $\beta_1 = \sqrt{-3}$ and $\beta_2 = -\sqrt{-3}$. Then we know that

$$N_3 = 3^1 + 1^1 - \beta_1^1 - \beta_2^1$$
$$N_9 = 3^2 + 1^2 - \beta_1^2 - \beta_2^2$$
$$N_{27} = 3^3 + 1^3 - \beta_1^3 - \beta_2^3$$

and so forth. The first equation only gives us information that we already had, but the second one predicts that $N_9 = 9 + 1 + 3 + 3 = 16$, exactly as computed in section 8 of chapter 8. Moreover, without doing any work at all, we can predict that $N_{27} = 28$. (You can check this claim if you have a *lot* of time on your hands.)

We can do this same magic trick (and it really is quite magical) for the curve $y^2 - y = x^3 - x$. We computed that $a_2 = -2$, so this time we know that $\zeta_{E,2}(T) = \frac{1 + 2T + 2T^2}{(1-T)(1-2T)}$. Now we factor $1 + 2T + 2T^2 = (1 - (-1 - i)T)(1 - (-1 + i)T)$. We have $\beta_1 = -1 - i$ and $\beta_2 = -1 + i$. Our formulas predict

$$N_2 = 2^1 + 1^1 - \beta_1^1 - \beta_2^1$$
$$N_4 = 2^2 + 1^2 - \beta_1^2 - \beta_2^2$$
$$N_8 = 2^3 + 1^3 - \beta_1^3 - \beta_2^3$$

and so forth. We can compute that $N_2 = 5$, $N_4 = 5$, $N_8 = 5$, $N_{16} = 25$, $N_{32} = 25$, $N_{64} = 65$, and $N_{128} = 145$. If you have even more time on your hands, you can verify some of these numbers.

5. Other L-Functions

We have explained a bit about how you find L-functions associated to **Z**-varieties. This construction can be generalized to other number systems besides **Z**, but we don't want to travel that road here. Instead, we want to mention briefly some other, very different, sources of L-functions. This section is not necessary for understanding the rest of the book, but all of these other L-functions are important for a deeper understanding of elliptic curves.

First, we have modular forms. A *cuspidal modular form* $f(q)$ is an infinite power series of the form

$$f(q) = \sum_{n \geq 1} b_n q^n$$

where the coefficients b_n are certain complex numbers. If you let $q = e^{2\pi i z}$, when z is a complex variable with $\Im(z) > 0$, then $f(q)$ becomes an analytic function with various stringent properties. It is those properties that make $f(q)$ a modular form. The set of all modular forms can be constructed from a subset of nicely normalized ones, called "newforms."

The *L*-function corresponding to $f(q)$ is defined by the Dirichlet series

$$L(f, s) = \sum_{n \geq 1} b_n n^{-s}.$$

If we assume that $f(q)$ is a newform, then there will be a finite set of bad primes and an Euler product:

$$L(f, s) = \text{fudge} \times \prod_{p \text{ good}} (1 - b_p p^{-s} + p^{k-1} p^{-2s})^{-1}$$

where k is a certain positive integer called the "weight" of the newform $f(q)$. If you compare this formula to $L(E, s)$, you'll see that it looks familiar.

In the next chapter, we will discuss the important analytic properties of *L*-functions. It turns out that these analytic properties have been proven for the *L*-function of any newform. The theorems of Wiles (1995), Taylor and Wiles (1995), and Breuil et al. (2001) say that for any elliptic curve E over **Q**, there exists a newform $f_E(q)$ of weight 2 with integer coefficients b_n such that $L(E, s) = L(f_E, s)$. We say that "every elliptic curve over **Q** is modular." This result is how you prove that $L(E, s)$ possesses all these nice analytic properties.

Another way of stating their theorem is that for any elliptic curve E over **Q**, there exists a newform $f(q)$ of weight 2 such that for every p, $a_p = b_p$, where the a_p's are the integers we've defined for E, and the b_p's are the coefficients appearing in the power series that defines $f(q)$. Readers of *Fearless Symmetry* will have seen how this theorem was used by Wiles to prove Fermat's Last Theorem.

A vast generalization of this theorem is only a conjecture, at the moment. It is part of the "Langlands program," which is a powerful set of conjectures and ideas which is motivating a large part of the research at the current frontiers of number theory. (Robert Langlands is a Canadian mathematician born in 1936.) First, you generalize the concept of modular form, but in such a way that you can still prove all the important analytic properties of their L-functions. These more general things are called *automorphic forms*. Then, you conjecture that any L-function attached to a **Z**-variety V is equal to the L-function of some automorphic form. This would then establish the nice analytic properties for these L-functions that go together to make up the Hasse–Weil zeta-function of V.

A third source of L-functions is *linear Galois representations*.[1] Let G be the absolute Galois group of **Q**. If K is a topological field, we call a continuous homomorphism $\rho : G \rightarrow \mathrm{GL}(n, K)$, a *linear Galois representation* of dimension n. (The notation $\mathrm{GL}(n, K)$ refers to the group of invertible $n \times n$-matrices with entries in K.) We assume ρ is unramified except for a finite set of primes p. We call that finite set the "bad primes" and the others the "good primes." If p is a good prime, then there is a well-defined conjugacy class of matrices denoted by $\rho(\mathrm{Frob}_p)$. Then we define the L-function of ρ by

$$L(\rho, s) = \text{fudge} \times \prod_{p \text{ good}} \det(I_n - p^{-s}\rho(\mathrm{Frob}_p))^{-1}.$$

If $n = 1$, ρ is usually called χ and $L(\chi, s)$ is Dirichlet's L-function we mentioned in section 1 of this chapter. Part of the Langlands program is the conjecture that for any ρ satisfying certain conditions, there should be an automorphic form whose L-function equals $L(\rho, s)$. If you could prove that, you might not get a million dollars right away, but you'd become immortal, maybe even get to be on TV, etc.

---

[1] This paragraph will only make sense to you if you read *Fearless Symmetry* or learned these concepts from somewhere else. When $n = 1$ and $K = \mathbf{C}$, you get Dirichlet's L-functions.

Chapter 14

· · · · ·

SURPRISING PROPERTIES OF *L*-FUNCTIONS

Road Map

Now that we have constructed the *L*-function associated to an elliptic curve, we focus our attention on the analytic properties of this function. We delay our consideration of how the *L*-function might capture some of the properties of the elliptic curve until the next chapter.

1. Compare and Contrast

The *L*-functions defined in the preceding chapter are not obviously functions of the complex variable s, although we told you they were. For example, let E be an elliptic curve defined over $\mathbf{Q}$. To state the BSD Conjecture, it is essential to view $L(E, s)$ as an actual function, not just a formal series, so that we can analytically continue it at least to the point where we can evaluate it at $s = 1$. As you will see in chapter 15, $s = 1$ is where all the action takes place.

We will see in this section that the series for $L(E, s)$ actually converges in some right half-plane (see figure 11.1 for an example), so that it defines an honest-to-goodness function there. Later in this chapter we will discuss the two most important properties of the function:

1. It analytically continues to the whole complex s-plane.
2. It satisfies an interesting functional equation.

Both of these properties were long-standing conjectures, finally resolved in Wiles (1995), Taylor and Wiles (1995), and Breuil et al. (2001).

There were certain kinds of elliptic curves for which these properties were already known, for example, those with "complex multiplication." We cannot discuss these kinds of curves further in this book, but they were very important in the history of understanding elliptic curves and making conjectures about them. Indeed, the data for the BSD Conjecture originally compiled by Birch and Swinnerton-Dyer were calculations using certain elliptic curves with complex multiplication.

As we discussed in the last chapter, there are many kinds of L-functions. The simpler ones of these, such as the Riemann zeta-function and Dirichlet's L-functions, were long known to have the two analytic properties listed above. In the early twentieth century, German mathematician Erich Hecke (1887–1947) showed that the L-functions associated to modular forms also had them. Presumably, this commonality was the motivation for the conjectures of Hasse, Japanese mathematician Yutaka Taniyama (1927–58), Japanese mathematician Goro Shimura (1930–), Weil, Langlands, and others, that various L-functions associated to $\mathbf{Z}$-varieties also have these analytic properties.

Recall that if S is the set of primes for which $E(\mathbf{F}_p)$ is singular, then

$$L(E, s) = \prod_{p \in S} \frac{1}{1 - a_p p^{-s}} \prod_{p \notin S} \frac{1}{1 - a_p p^{-s} + p^{1-2s}}.$$

where the a_p's were defined in section 4 of the preceding chapter. Remember also that we can rewrite the product for $L(E, s)$ as a Dirichlet series:

$$L(E, s) = \sum_{n \geq 1} \frac{a_n}{n^s} = \sum_{n \geq 1} a_n n^{-s}.$$

The a_n's are all determined, of course, by the a_p's.

In chapter 11, we stated some theorems about convergence of Dirichlet series. You can take these theorems on faith, but we would like to review here the basic reason from calculus why they work. Let's suppose we have any Dirichlet series

$$f(s) = \sum_{n \geq 1} a_n n^{-s}$$

where $s = \sigma + it$ is a complex variable. As we saw on p. 168,

$$n^{-s} = e^{-\sigma \log n} e^{-it \log n}.$$

For any real number y, we know the formulas $e^{iy} = \cos(y) + i\sin(y)$ and $\cos^2 y + \sin^2 y = 1$. Together they tell us about the absolute value: $|e^{iy}| = 1$. In particular this is true of $e^{-it \log n}$. Now $|zw| = |z||w|$. In conclusion we have the all-important fact that

$$|n^{-s}| = |e^{-\sigma \log n}| = |n^{-\sigma}|.$$

We say that a series $\sum_{n=1}^{\infty} b_n$ *converges absolutely* if and only if the series of positive real numbers $\sum_{n=1}^{\infty} |b_n|$ converges. It is a theorem that if a series converges absolutely, it also converge in the plain vanilla sense. In other words, the partial sums $\sum_{n=1}^{k} b_n$ get closer and closer to some limiting value L as k increases, and in fact approach and stay as close as you like to L if you take k big enough.

Let's apply these ideas to a Dirichlet series, so that $b_n = a_n n^{-s}$. Then $|b_n| = |a_n||n^{-\sigma}|$, and the series will converge absolutely if and only if $\sum_{n=1}^{\infty} |a_n||n^{-\sigma}|$ converges. Aha! The imaginary part it of s has disappeared from the formula. That means that if the Dirichlet series converges absolutely for one value of s with the real part σ, it will converge for *all* s with the same real part σ. All those s's form a vertical line, above and below σ on the real axis.

Next, we can invoke the "comparison test" about series, which you may have learned in calculus. This very plausible theorem says that if you have a convergent series of positive numbers $\sum_{n=1}^{\infty} c_n$ that converges to L, say, and if you have another series $\sum_{n=1}^{\infty} b_n$ of any old complex numbers and if you know

$$|b_n| \leq c_n$$

for all n (or even just for all n greater than some fixed N), then $\sum_{n=1}^{\infty} b_n$ converges absolutely. This is common sense, because the partial sums $\sum_{n=1}^{k} |b_n|$ are increasing as k increases, and haven't got a lot of room to wiggle in, since they are all bounded by the positive number L.

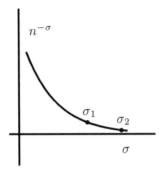

Figure 14.1. $n^{-\sigma}$

Let's compare the Dirichlet series $\sum_{n\geq 1} a_n n^{-s}$ for two values of s, one with real part σ_1 and one with real part σ_2, where $\sigma_1 < \sigma_2$. Now $n^{-\sigma}$ is monotonically decreasing as σ increases. In fact, it is dying off exponentially, as you can see in figure 14.1. This means that for every n, $|a_n||n^{-\sigma_2}| < |a_n||n^{-\sigma_1}|$. So suppose you know that the Dirichlet series is absolutely convergent when the real part of s is σ_1. The comparison theorem then tells you that it is absolutely convergent when the real part of s is σ_2.

This means that if the Dirichlet series is absolutely convergent for the values of s on one vertical line, it is absolutely convergent for the values of s on any vertical line to the right of the first one. When you pile these assertions all together, you get a right half-plane. This is the basic reason that Dirichlet series converge absolutely in some right half-plane.

An important theorem from complex analysis tells us that if a Dirichlet series converges absolutely in some open right half-plane, then the function it converges to (in the vanilla sense) is an analytic function of s. Let's apply this theorem to the Dirichlet series for $L(E, s)$. In equation (13.3), we saw that

$$\frac{1}{1 - a_p p^{-s}} = 1 + a_p p^{-s} + a_p^2 (p^2)^{-s} + a_p^3 (p^3)^{-s} + \cdots$$

and equation (13.4) said that

$$\frac{1}{1 - a_p p^{-s} + p^{1-2s}} = 1 + a_p p^{-s} - p(p^2)^{-s} + a_p^2 (p^2)^{-s} + \cdots .$$

From these formulas, and the bound $|a_p| \leq 2\sqrt{p}$ proved by Hasse, we can work out a similar bound for every a_n. Now we can't get $|a_n| \leq 2\sqrt{n}$ on the nose. It's a little worse than that, but not too bad.

For example, suppose that p and q are two *different* primes, and let $n = pq$. Where could a term $a_n n^{-s}$ come from in the Euler product? The only term that contains p^{-s} is $a_p p^{-s}$ and the only term that contains q^{-s} is $a_q q^{-s}$. The only way we get n is by multiplying p and q. (See again how important the unique factorization theorem is in number theory.) So the coefficient a_n of n^{-s}, when we multiply together all the Euler factors, is going to be $a_p a_q$. Now $|a_p| \leq 2\sqrt{p}$ and $|a_q| \leq 2\sqrt{q}$, so it follows that $|a_n| = |a_p a_q| = |a_p||a_q| \leq (2\sqrt{p})(2\sqrt{q}) = 4\sqrt{pq} = 4\sqrt{n}$.

The original factor of 2 has become a factor of 4, which is not good for us: You can see that if n has many different prime factors, the 2 is going to become an even larger power of 2. So we can't find any constant K so that $|a_n| \leq K\sqrt{n}$. However, we can quash all these constants by increasing the exponent on n a bit.

The analysis gets somewhat complicated and we won't continue in detail. But just to get the flavor, suppose the Dirichlet series contained only the a_n's for n prime. Then using the bound $|a_p| \leq 2\sqrt{p}$, we would conclude that the Dirichlet series for $L(E, s)$ would converge absolutely for any $s = \sigma + it$ as long as $\sum_{n \geq 1} n^{1/2} n^{-\sigma} = \sum_{n \geq 1} n^{1/2-\sigma}$ converges.

In fact, this turns out to be the correct result, even if you include all the terms of the Dirichlet series. So we repeat, now in a more authoritative voice:

THEOREM 14.1: The Dirichlet series for $L(E, s)$ will converge absolutely for any $s = \sigma + it$ as long as $\sum_{n \geq 1} n^{1/2-\sigma}$ converges.

So what σ's will produce convergence here? We have to dip back into calculus. The "integral test" tells you that $\sum_{n \geq 1} \frac{1}{n^x}$ converges if $x > 1$. That statement is exact: The sum diverges if $x \leq 1$. Therefore, $\sum_{n \geq 1} n^{1/2-\sigma}$ converges if and only if $1/2 - \sigma < -1$. This happens when $\sigma > 3/2$. In conclusion:

THEOREM 14.2: The Dirichlet series for $L(E, s)$ converges absolutely in the right half-plane $\Re(s) > 3/2$.

2. Analytic Continuation

We have seen in the preceding section that $L(E, s)$ converges absolutely and therefore defines an analytic function in the right half-plane $H = \{s \in \mathbf{C} \mid \Re(s) > 3/2\}$. If you go back now to chapter 12, you will see that there is the possibility that this analytic function $L(E, s)$ might be extendable analytically to a larger domain than H. If so, this extension may tell us a lot.

First of all, it means that the a_p's are not some random collection of integers. If they didn't cohere in a powerfully intrinsic way, the Euler product defined using a_p would probably not have any nice properties at all, beyond absolute convergence in the original domain H. There is no reason the function should analytically continue to a larger domain.

Second, if you peek ahead at chapter 15, you will see that it is essential for us to evaluate $L(E, s)$ at and near $s = 1$ to formulate the BSD Conjecture. You will also see why it is exactly the point $s = 1$ that pops up in the conjecture. But $s = 1$ isn't in the half-plane H. So without knowing analytic continuation, we couldn't even begin to make the conjecture. (Well, that's not exactly true. In the 1960s, when Birch and Swinnerton-Dyer made their conjecture, analytic continuation was only known for elliptic curves with complex multiplication and for modular elliptic curves. It was later proved that all elliptic curves are modular.)

Third, another form of coherence of the a_p's is given by a functional equation, which we will discuss in the next section. The functional equation relates $L(E, s)$ and $L(E, 2 - s)$, so the functional equation wouldn't make much sense without analytically continuing $L(E, s)$ to be a function for all s, or at least for some s's with real part smaller than $3/2$.

Fourth, it is expected that knowledge of where $L(E, s)$ equals zero to the left of H will be important. This is related to what is called the *Generalized Riemann Hypothesis*. To formulate this hypothesis, again we need to analytically continue $L(E, s)$.

Fifth, there is an amazing theorem that the Dirichlet series for $L(E, s)$ actually converges (not absolutely) for any s with real part greater than $5/6$. However, this theorem can only be proved by the same methods that prove the analytic continuation and functional equation anyway. By the way, you might think that once you know that the Dirichlet series for $L(E, 1)$

actually converges, you might be able to use it to compute $L(E, s)$ and its derivatives at $s = 1$ and maybe to prove the BSD Conjecture. But this is extremely unlikely. Convergent series that are not absolutely convergent are very hard to handle, and usually converge very slowly, so they are not even good for computational purposes.

In fact, $L(E, s)$ can be continued to be an analytic function for all values of s:

> **THEOREM 14.3:** Let E be an elliptic curve defined over **Q**. Then the function $L(E, s)$, which we know is an analytic function in the right half-plane H, actually continues analytically to be an analytic function in the entire s-plane.

> **PROOF:** The only known way of proving this theorem is to show that E is modular. That means to show that there is some modular form $\sum a_n q^n$ such that the a_n's are the same as the a_n's in the Dirichlet series for $L(E, s)$. In addition to knowledge of modular forms, the proof requires knowing

> 1. that the L-function of a modular form has analytic continuation (and a functional equation), and, as we said,
> 2. that every E is modular.

> Point 1 was covered by Hecke in the early twentieth century, but it requires a lot of advanced mathematics to explain. Point 2 is the main outcome of the work of Wiles (1995), Taylor and Wiles (1995), and Breuil et al. (2001), which we keep mentioning. It requires a lot more advanced mathematics to explain, and we can't discuss it further in this book.

3. Functional Equation

The other major property of $L(E, s)$ is a functional equation. Once we know that $L(E, s)$ continues to an analytic function for all s—and we will assume theorem 14.3 from now on—we can ask what additional properties $L(E, s)$ has.

A functional equation should relate the values of a function in one domain to its values in a separate domain. Consider a simple example.

We can define e^x as a function for real x. It analytically continues to be an analytic function of z for every z in the complex plane. (The proof of this is not too hard. The function e^x is given by its Taylor series, and when we plug in any complex z at all instead of x, the Taylor series converges absolutely, and so defines an analytic function for all z.) There is then a functional equation:

$$e^{-z} = \frac{1}{e^z}.$$

In this example, the functional equation relates the values of the exponential function at complex numbers with positive real part to the values with negative real part. For more about functional equations, look back at section 4 of chapter 11.

This example has another feature: Because the values in a right half-plane are related to the values in a left half-plane, there may be a line where the half-planes come together. In the case of e^z, it is the imaginary axis (where z has 0 for its real part). This is called the "critical line" for the function. The functional equation relates the value at a point on the critical line to a value at some other point on the critical line.

The values of a function on its critical line are often quite interesting. For example, the critical line of the Riemann zeta-function is $\Re(s) = \frac{1}{2}$, and the Riemann Hypothesis concerns that line especially.

The functional equation for $L(E, s)$ will relate $L(E, s)$ to $L(E, 2 - s)$.

EXERCISE: What will be the critical line for this functional equation?

SOLUTION: The critical line will occur when s and $2 - s$ have the same imaginary part. If we call that imaginary part y, then we must have $y = 2 - y$. That's an equation even we can solve: $y = 1$ is the solution. So the critical line for $L(E, s)$ is the line given by $\Re(s) = 1$. Hey, look! The crucial point that is so important for us, $s = 1$, lies on the critical line. This is not a coincidence—it is a phenomenon that seems to come up often in the study of L-functions, as we just mentioned.

It turns out that to obtain the functional equation for $L(E, s)$, we will have to throw in some new factors made out of gamma-functions, which we introduced in chapter 11. Remember that $\Gamma(s)$ is analytic for all s, except for $s = 0, -1, -2, \ldots$. We form a new function by multiplying $L(E, s)$ by a gamma-function (following the same process as when writing down the functional equation of the Riemann zeta-function) thus:

$$\Lambda(E, s) = \frac{(\sqrt{N})^s}{(2\pi)^s}\Gamma(s)L(E, s).$$

As you can see we also threw in a couple of other exponential factors.

One of the exponential factors appearing in the functional equation is the s-th power of the square root of a positive integer N. What is N? It is a number that depends on the particular elliptic curve E, and it is called the *conductor* of E. It could be defined as the number whose square root raised to the s-th power shows up in the definition of $\Lambda(E, s)$, though of course that definition is a tautology. The conductor is an integer whose prime factors are taken from the prime factors of Δ_E, and usually those prime factors appear in the factorization of N to a lower power than they do in the factorization of Δ_E. There are formulas for computing N from the cubic equation that defines E.[1]

THEOREM 14.4: The function $\Lambda(E, s)$ is analytic in the whole s-plane, and satisfies the functional equation

$$\Lambda(E, 2 - s) = w\Lambda(E, s)$$

where w is either 1 or -1.

The proof of theorem 14.4, as the proof of analytic continuation in the preceding section, involves showing that E is modular, and then knowing that the L-function of a modular form has the same properties.

The functional equation (via the L-function embedded in $\Lambda(E, s)$) again shows us there is some very strong and mysterious coherence in the a_p's coming from the elliptic curve E. We won't be using the functional equation in the rest of this book, except that right now we can notice something interesting about the number w.

---

[1] We have asserted several times that every elliptic curve E defined over **Q** is *modular*. See p. 213. Every modular form has a positive integer attached to it, called its *level*. It turns out that the level of $f_E(q)$ and the conductor of E are always equal.

The number w can be computed for any given E in a way that we will not describe. Now let's look at the Taylor series for $\Lambda(E, s)$ at $s = 1$:

$$\Lambda(E, s) = c(s - 1)^\rho + \text{h.o.t.}$$

where $c \neq 0$ and h.o.t., standing for "higher order terms," means the infinite sum of all the terms involving the powers higher than ρ.

The extra factors we threw in to go from L to Λ are $\frac{(\sqrt{N})^s}{(2\pi)^s}$ and $\Gamma(s)$. When $s = 1$, they are nice, nonzero, numbers: They evaluate to $\frac{\sqrt{N}}{2\pi}$ and 1, respectively. So, at $s = 1$, the leading exponent ρ for $\Lambda(E, s)$ is the same as the leading exponent for the Taylor series of $L(E, s)$. In other words, ρ is the "analytic rank" of E, as defined later, in chapter 15.

Now let's do the functional equation to the Taylor series:

$$\Lambda(E, 2 - s) = c((2 - s) - 1)^\rho + \text{h.o.t.} = c((1 - s))^\rho + \text{h.o.t.}$$

$$= uc(s - 1)^\rho + \text{h.o.t.}$$

where $u = 1$ if ρ is even and $u = -1$ if ρ is odd. But by the functional equation,

$$\Lambda(E, 2 - s) = w\Lambda(E, s) = wc(s - 1)^\rho + \text{h.o.t.}$$

We conclude that $u = w$. So if $w = 1$, then ρ is even, and if $w = -1$, then ρ is odd.

As we will see, the BSD Conjecture says that ρ is also the algebraic rank (to be defined later, in chapter 15) of E. So the BSD Conjecture predicts that knowledge of w will tell you whether the algebraic rank of E is odd or even. This is a much simpler statement than the full BSD Conjecture, and people have made some progress toward proving it. In fact, if a certain group called the Tate–Shafarevich group of E is finite, then it can be proved that $w = 1$ if and only if the algebraic rank of E is even. For more details on this and much else that we can only hint at in this book, see Rubin and Silverberg (2002).

Chapter 15

. . . . .

THE CONJECTURE OF BIRCH AND SWINNERTON-DYER

Road Map

We are finally at our journey's end. Armed with our tools, we can describe the mysterious conjecture we have aimed at all along. It relates properties of the L-function to properties of the elliptic curve used to construct the L-function.

The conjecture says that the order of the zero of the L-function of an elliptic curve E at $s = 1$ is equal to the rank of the abelian group $E(\mathbf{Q})$ of rational points on the elliptic curve. In this way, the number of rational solutions to a cubic equations with integer coefficients can be connected to analytic properties of generating functions coming from the solutions of the same equation over all the finite fields.

1. How Big Is Big?

In this chapter, we will look at cubic equations $y^2 = x^3 + Ax + B$ where A and B are integers, and we will ask: How many rational solutions are there to this equation? Answering this question is the goal we've been aiming at all along.

Let's fix A and B. Then we are looking for rational numbers x and y solving that equation: $y^2 = x^3 + Ax + B$. At first, that doesn't seem so hard. All we need to solve it is to find a rational number x such that $x^3 + Ax + B$ is a perfect square (which means that if you write it as a fraction in lowest terms, the numerator and denominator both have to be perfect squares).

This situation is very typical of Diophantine number theory. Diophantus himself has problems such as: Make such-and-such an expression in an unknown rational number into a perfect square. These sorts of problems are the bedrock of algebraic number theory. How well we do at solving them is a good indicator of how well we understand the hidden structures of numbers and their relationships.

We'd like to find such x, and while we're at it, we might as well try to find *all* such x. Now, for a particular choice of A and B, these tasks can often be accomplished, but there is no guarantee. That is to say, there is no general algorithm known which will find all, or even *any* rational solutions to a given equation $y^2 = x^3 + Ax + B$. "General" means a single algorithm that will work for any A and B as inputs. The known techniques for solving particular cubic equations are beyond the scope of this book, and can be found in various text books and research papers. For example, you can look at the beautiful expository paper (Rubin and Silverberg, 2002), as well as the references in that paper.

Even if we had a general algorithm, it might not shed that much light on the generality of behavior to be expected as we vary A and B. Rather than ask for specific solutions (which will generally be fractions x and y with huge numerators and denominators, and not so illuminating), we can ask more qualitative questions.

For example, if you pick an A and a B, are there usually "lots" of solutions? How can we measure the size of the number of solutions? Can we predict this size? We've been laboring in this book to provide the background needed to give a precise way to measure this size and explain the Conjecture of Birch and Swinnerton-Dyer about this size.

Birch and Swinnerton-Dyer made their conjecture in the 1960s. We will explain what data led them to make their conjecture. For a good very brief account of the conjecture (written for professional mathematicians), including some of the history, see Wiles (2006). This account, as does Rubin and Silverberg (2002), includes some indication of what progress has been made toward proving the conjecture. We will later mention one theorem in this regard.

The BSD Conjecture has a strong form and a weak form. In this book, we explain the weak form, and tell you what the strong form is.

Fix integers A and B. We shall assume that the cubic $x^3 + Ax + B$ has 3 distinct roots. This assumption implies that the complex projective curve E of degree 3 defined by the homogeneous polynomial equation $y^2z - x^3 - Axz^2 - Bz^3 = 0$ is nonsingular. We say that E is "defined over $\mathbf{Q}$" because A and B are rational numbers. We have seen in previous chapters that if K is *any* field, the nonsingular points of $E(K)$ form a group.

Instead of talking about solutions to $y^2 = x^3 + Ax + B$, we will talk about points on $E(K)$. Because of the point at infinity, there will be one more point on $E(K)$ than solutions to $y^2 = x^3 + Ax + B$. Actually, we work with the "minimal model" for E, which may be more complicated than $y^2 = x^3 + Ax + B$, because we must be prepared to work over $\mathbf{F}_p$ for *all* primes p, including $p = 2$ and $p = 3$. See the first paragraph of section 4 in chapter 13.

Our main question now becomes: "How many points are there in $E(\mathbf{Q})$?" Crudely, we could say there are either finitely many or infinitely many. But we have learned a more refined way of discussing this. The group $E(\mathbf{Q})$ is a finitely generated abelian group. So it has a finite torsion subgroup, $E(\mathbf{Q})_{\text{tors}}$ and it has a rank $r \geq 0$. The group $E(\mathbf{Q})$ is finite if and only if $r = 0$. But when the group is infinite, we can sophisticatedly distinguish between "sizes" of infinity by asking about the value of r.

This number r is called the "algebraic rank of E." It is the number that measures how many independent nontorsion solutions there are to our cubic equation. In the realm of Diophantine number theory, it is the main qualitative information we could seek about the cubic equation. It is so important that we will repeat the definition:

DEFINITION: Let E be an elliptic curve defined over $\mathbf{Q}$. The *algebraic rank* of E is the rank r of the finitely generated abelian group $E(\mathbf{Q})$.

Another way of making this definition would be to say that the rank of E is the largest integer r such that there are r "independent" points $P_1, \ldots, P_r$ in $E(\mathbf{Q})$.

DEFINITION: Let A be an abelian group and $a_1, \ldots, a_k$ elements of A. Then $a_1, \ldots, a_k$ are *independent* if and only if the equation

$$n_1 a_1 + \cdots + n_k a_k = 0$$

with integers $n_1, \ldots, n_k$ implies that $n_1 = \cdots = n_k = 0$.

For example, E has rank 0 or 1 under the following conditions:

1. E has rank 0 if $E(\mathbf{Q})$ is a finite group. Thus if P is any element of $E(\mathbf{Q})$, either $P = \mathcal{O}$ or $P + P + \cdots + P = \mathcal{O}$ for some number of P's.
2. E has rank 1 if $E(\mathbf{Q})$ contains infinitely many elements, and, for any two elements P and Q of $E(\mathbf{Q})$, you can always find integers a and b, not both zero, such that $aP + bQ = \mathcal{O}$. That is to say, either P has finite order n (in which case you can take $a = n$, $b = 0$), or Q has finite order m (in which case you can take $a = 0$, $b = m$), or both P and Q have infinite order, and either

$$\overbrace{P + P + \cdots + P}^{a} = \overbrace{Q + Q + \cdots + Q}^{|b|} \text{ or } \overbrace{P + P + \cdots + P}^{a} =$$

$$\underbrace{(-Q) + (-Q) + \cdots + (-Q)}_{|b|}, \text{ where there are } a \text{ } P\text{'s and } |b| \text{ } Q\text{'s or}$$

$-Q$'s added up.

We've explained this in such a long-winded way so that you can see that the rank of E has to do with what happens when you start adding up various points on $E(\mathbf{Q})$, which means connecting two points by a line, seeing where the line meets the curve E at a third point, reflecting in the x-axis and doing this madly again and again and seeing if you ever end up at $\mathcal{O}$, that is, at the point at infinity. In general: E has rank $r \geq 1$ if

1. You can find r points $P_1, \ldots, P_r$ in $E(\mathbf{Q})$.
2. No matter how you add multiples of these points or their negatives over and over, you don't get $\mathcal{O}$, unless you do it some trivial way.
3. If you take $r + 1$ points, you can always find a nontrivial way of adding them up to get $\mathcal{O}$.

So the rank of an elliptic curve over $\mathbf{Q}$ measures "how big" the group of rational points $E(\mathbf{Q})$ is. And this in turn measures "how many" rational solutions there are to the cubic equation $y^2 = x^3 + Ax + B$ that defines E.

2. Influences of the Rank on the N_p's

A basic insight (or guess) of Birch and Swinnerton-Dyer was that the amount of rational solutions to $y^2 = x^3 + Ax + B$ would be reflected in the number of solutions modulo p of that equation for various primes p, and conversely. To quote Birch and Swinnerton-Dyer (1963):

> ...one hopes that if for most p the curve [given by the equation $y^2 = x^3 + Ax + B$, where A and B are rational] has unusually many points in the finite field with p elements, then it will have a lot of rational points.

The point here is that you can not only reduce integer solutions modulo p to get solutions over $\mathbf{F}_p$, you can even reduce merely rational solutions modulo p (usually). To understand this, suppose $x = a/b$ and $y = c/d$ are two rational numbers satisfying the equation $y^2 = x^3 + Ax + B$. Let p be a prime number and assume that neither b nor d are divisible by p. Then you can reduce x and y modulo p and get a solution to the congruential equation $y^2 \equiv x^3 + Ax + B \pmod{p}$.

Let's see how this works in an example. Suppose $x = -8/9$ and $y = 109/27$. They satisfy[1] the equation $y^2 = x^3 + 17$. Choose the prime $p = 5$. The equation becomes $y^2 \equiv x^3 + 2 \pmod 5$. Now $x \pmod 5$ becomes $-3/(-1) \equiv 3$. Meanwhile $y \pmod 5$ becomes $-1/2 \equiv 2$ because the multiplicative inverse of 2 is 3 (because $2 \cdot 3 \equiv 1 \pmod 5$) so $-1/2 \equiv (-1) \cdot 3 \equiv -3 \equiv 2$. So $x = 3, y = 2$ is a solution[2] to $y^2 \equiv x^3 + 2 \pmod 5$.

Instead, let's choose the prime $p = 7$.

EXERCISE: Work out x and $y \pmod 7$ and check they solve $y^2 \equiv x^3 + 17 \pmod 7$.

SOLUTION: The equation becomes $y^2 \equiv x^3 + 3 \pmod 7$. Now $x \pmod 7$ becomes $-1/2 \equiv 3$. Meanwhile $y \pmod 7$ becomes $4/(-1) \equiv 3$. So $x = 3, y = 3$ is a solution[3] to $y^2 \equiv x^3 + 2 \pmod 7$.

And so it goes. For every prime p not equal to 3, we can reduce $x = -8/9, y = 109/27 \pmod p$ and get a solution to $y^2 \equiv x^3 + 17 \pmod p$.

---

[1] CHECK: $x^3 = -512/729$ so $x^3 + 17 = -512/729 + 17 \cdot 729/729 = (12393 - 512)/729 = 11881/729 = (109^2)/729 = y^2$.

[2] CHECK: $x^3 \equiv 27 \equiv 2$, so $x^3 + 2 \equiv 4 \equiv 2^2 \equiv y^2$. Notice how much easier it is to do arithmetic modulo 5 than with fractions.

[3] CHECK: $x^3 \equiv 27 \equiv -1$, so $x^3 + 3 \equiv 2 \equiv 9 \equiv 3^2 \equiv y^2$.

Thus a single point on $E(\mathbf{Q})$ yields a point on $E(\mathbf{F}_p)$ for all but a finite number of p's.

Remember that N_p is the number of points in $E_{\mathrm{ns}}(\mathbf{F}_p)$. For all but the finitely many p that are factors of the minimal discriminant Δ_E of E, E modulo p is nonsingular, so N_p is simply the number of points in $E(\mathbf{F}_p)$.

We also defined the integers a_p. If E modulo p is nonsingular, we have

$$N_p = p + 1 - a_p,$$

while if E modulo p is singular, the details are in table 9.4.

Thus a point P on $E(\mathbf{Q})$ will contribute 1 to the count for N_p, for each but a finite number of p's. If P is not $\mathcal{O}$, then adding it to itself creates more points on $E(\mathbf{Q})$: $2P, 3P, 4P, \ldots$, and their negatives give even more points: $-P, -2P, -3P, \ldots$. Of course, modulo p some of these points will have to start coinciding, because $E(\mathbf{F}_p)$ is only a finite set.

Now if P is a point of finite order in $E(\mathbf{Q})$, these multiples of P start repeating, and they will yield a fixed number at most in the count of N_p for various p's. For example, if P has order 6, then P and its multiples can bump up the count of N_p at most by 5. Since we always have the neutral element $\mathcal{O}$, we always have $N_p \geq 1$, and a point of order 6 would bump up the count maximally only if P, $2P$, $3P$, $4P$, and $5P$ were all distinct modulo p. (Don't forget $6P = \mathcal{O}$.)

As p get large, a fixed bump-up like this becomes negligible in the count, since N_p is pretty close to $p + 1$. (By the bound proven by Hasse, we know $|p + 1 - N_p| < 2\sqrt{p}$.)

So if we want significant contributions to the N_p's for all p's, we had better use a point P of infinite order. Then we can hope that modulo p, the set of all multiples of P will spread out in $E(\mathbf{F}_p)$ and contribute a lot of distinct solutions modulo p. If we have *two* independent points of infinite order P and Q, then we have all their linear combinations $\{aP + bQ\}$ with integers a and b, and we can expect they will spread out even more and contribute even more to the size of N_p. And so on for larger and larger algebraic ranks.

How shall we measure the size of the N_p's? A good measure is the following. First, since $N_p = p - a_p + 1$ and the largest term in this sum is p, we should "normalize" by considering N_p/p instead of N_p itself. This puts all the primes p on a similar footing. No matter what p is, N_p/p won't be far off from 1, and in fact it will always be between $1 - \frac{2}{\sqrt{p}} + \frac{1}{p}$ and $1 + \frac{2}{\sqrt{p}} + \frac{1}{p}$. (We got this range by starting with the inequalities $-2\sqrt{p} < a_p < 2\sqrt{p}$ and dividing through by p.) The more negative a_p, the larger N_p, and vice versa.

OK, what now? If we just look at the limit of N_p/p as p tends to ∞, we'll just get 1, no matter what. Not very illuminating. Instead, let's add up all the N_p/p's and see what we get. Hmm, we get ∞, since we are basically adding up a bunch of 1's. Even if we look at the partial sums and see how fast they are going to infinity, the answer won't depend on E: They will be growing like the partial sums of $1 + 1 + 1 + \cdots$.

The idea of Birch and Swinnerton-Dyer is look at partial *products* of the N_p/p's and see *how fast* they are getting big. We define a function

$$\beta_E(x) = \prod_{p \leq x} \frac{N_p}{p}.$$

(The notation $\prod_{p \leq x}$ just means taking the product of all terms N_p/p for all primes up to the real number x.) Now the product of a bunch of 1's is just 1. So the infinite product of a bunch of numbers all of which are around 1 is a rather subtle thing. It might have a finite limit, or it might tend to infinity, or it might tend to 0, or it might have no limit at all.[4]

Birch and Swinnerton-Dyer were led to compare the growth of $\beta_E(x)$ as $x \to \infty$ with the growth of powers of the logarithm of x. In their computer experiments, they graphed the function $\log \beta_E(x)$ against $\log \log(x)$, and saw that in each case, the graph, although jagged, seemed to stay near a line with a certain slope. And that slope was the algebraic rank of E! Examples of these graphs may be conveniently seen in Rubin and Silverberg (2002).

To make this more precise, we introduce the twiddle notation. If you have two functions of the real variable x, say $f(x)$ and $g(x)$, then we write $f(x) \sim g(x)$ if and only if the limit of $f(x)/g(x)$ as $x \to \infty$ exists and equals 1. We can now state the basic BSD observation like this:

---

[4] We can take logarithms, and relate this limit to the behavior of the infinite sum of the logarithms of the numbers. This procedure becomes reminiscent of the definition of the Hasse–Weil zeta-function, which is partially why Birch and Swinnerton-Dyer were eventually led to expressing their conjecture in terms of L-functions.

Their jagged graphs led Birch and Swinnerton-Dyer to guess that if E is an elliptic curve over $\mathbf{Q}$ of algebraic rank r, then

$$\beta_E(x) \sim C(\log x)^r \tag{15.1}$$

for some nonzero, positive constant C.

Birch and Swinnerton-Dyer also had some ideas about what C should be equal to. As we shall see later in this chapter, they were led to make their famous conjecture in a somewhat different form, stated below as conjecture 15.3, along with equation (15.4). Equation (15.1) implies the conjecture, that is, if (15.1) is true, then conjecture 15.3 is also true. (This sentence depends on a theorem in Goldfeld (1982), plus the fact that every elliptic curve is "modular," which was proved by the combined work in Wiles (1995), Taylor and Wiles (1995) and Breuil et al. (2001).) However, (15.1) is known to be stronger than the conjecture, that is, it implies more things than conjecture 15.3, things that are currently not known to be true. At the moment, there seems to be no consensus about whether (15.1) is likely to be true or not.

We pause here to emphasize the experimental way in which Birch and Swinnerton-Dyer arrived at their conjecture. Many parts of mathematics, both pure and applied, are amenable to experimental methods that help to find and refine assertions that one hopes to prove, sooner or later. This occurs at high levels, like the BSD Conjecture, and also at very low levels, for example when trying to prove little lemmas in the course of a larger project. The advent of electronic computers made such experimentation much more widespread, but it has always been a part of mathematics. Number theory is especially amenable to experimentation, but abstract algebra, differential geometry, topology, and many other parts of mathematics are now routinely aided by experimental methods using computers.

3. How Small Is Zero?

Suppose you have a function $f(t)$ that equals zero at some point. Let's say $f(1) = 0$. How small is that 0? Stupid question? Not necessarily. What we are really asking, is *how fast* does $f(t)$ tend to 0 as t tends to 1. For

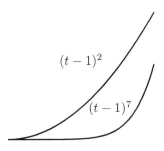

Figure 15.1. $(t-1)^7$ vs. $(t-1)^2$

instance, $(t-1)^7$ goes to zero much faster than $(t-1)^2$, as you can see in figure 15.1. The higher the exponent, the faster we get to zero.

There can be rates of tending to 0 which are not given by integer exponents. For example, $(t-1)\sqrt{t-1}$ goes to zero as $t \to 1$ somewhat slower than $(t-1)^2$ but somewhat faster than $(t-1)$. But this function is not differentiable infinitely many times at $t = 1$, because the second derivative of $(t-1)\sqrt{t-1}$ is $\frac{3}{4\sqrt{t-1}}$, which is not defined at $t = 1$. Therefore, it is not defined by a convergent Taylor series at $t = 1$.

Let's restrict our attention to analytic functions. As we've discussed in chapter 12, that means that around any point where the function is defined, it has an infinite Taylor series that determines the function in some open disc centered at that point. Also, analytic functions are functions of a complex variable, not just a real variable.

Let $f(s)$ be an analytic function defined in an open disc around $s = 1$. Then we can write down its Taylor series around 1:

$$f(s) = k(s-1)^m + k_1(s-1)^{m+1} + k_2(s-1)^{m+2} + \cdots .$$

The Taylor series begins with the product of some nonzero constant k and $(s-1)^m$ for some $m \geq 0$. We have seen that the exponent m is called the *order of the zero* of $f(s)$ at $s = 1$, and the nonzero number k is called the *leading coefficient* of $f(s)$ at $s = 1$. Thus m measures how fast $f(s) \to 0$ as $s \to 1$.

We can apply these ideas to the L-function of an elliptic curve E defined over $\mathbf{Q}$, namely to $L(E, s)$. As we have seen, $L(E, s)$ is initially defined by a series of functions of s that converges absolutely for $\Re(s) > \frac{3}{2}$. But it has been proven that $L(E, s)$ has an analytic continuation to the whole s-plane.

In particular, it has a Taylor series expansion at $s = 1$:

$$L(E, s) = c(s - 1)^\rho + k_1(s - 1)^{\rho+1} + k_2(s - 1)^{\rho+2} + \cdots,$$

where $c \neq 0$. So ρ tells us how fast $L(E, s) \to 0$ as $s \to 1$. We will soon see that this is related to how fast $\beta_E(x) \to \infty$ as $x \to \infty$. Since this last order of growth is related to the algebraic rank of E, it is traditional also to call ρ a rank, the "analytic rank" of E.

> **DEFINITION**: Let E be an elliptic curve defined over **Q**. The *analytic rank* of E is ρ, the order of the zero of $L(E, s)$ at $s = 1$.

If we go back to the definition of $L(E, s)$ and play some crazy (or creative) games with it, we can see how ρ could also be related to the N_p's for E. Of course, in some sense it is very related, because if you know all the N_p's, and you know exactly for which p's $E(\mathbf{F}_p)$ is singular, you can reconstruct the function $L(E, s)$ on the nose. (Even if you don't know which $E(\mathbf{F}_p)$'s are singular, it doesn't affect the computation of ρ, because ρ won't change if you change a finite number of factors in the Euler product for $L(E, s)$.)

First, from the N_p's, you get the a_p's from the formulas above. Let S be the set of primes for which $E(\mathbf{F}_p)$ is singular. Then you form the L-function:

$$L(E, s) = \prod_{p \in S} \frac{1}{1 - a_p p^{-s}} \prod_{p \notin S} \frac{1}{1 - a_p p^{-s} + p^{1-2s}}.$$

As explained earlier, the infinite (second) product is certain to converge only when $\Re(s) > 3/2$. The magic of analytic continuation extends the L-function in some difficult way to have values for all complex s. But just for fun, let's simply set $s = 1$ and see what we get:

$$\text{"}L(E, 1)\text{"} \quad \text{"} = \text{"} \quad \text{"} \prod_{p \in S} \frac{1}{1 - a_p p^{-1}} \prod_{p \notin S} \frac{1}{1 - a_p p^{-1} + p^{-1}}.\text{"}$$

We put everything in scare quotes because the infinite product on the right doesn't really mean anything. It's interesting that all the exponents on the p's now become -1.

We can replace the right-hand side with partial products, up to $p \leq x$ for any real number x greater than the largest prime in S. When we do, we get

$$\text{RHS up to } x = \prod_{p \in S} \frac{1}{1 - a_p p^{-1}} \prod_{p \notin S, p \leq x} \frac{1}{1 - a_p p^{-1} + p^{-1}}.$$

We can take off the quotes because now we are dealing with finite products. Taking reciprocals gives

$$\frac{1}{\text{RHS up to } x} = \prod_{p \in S} (1 - a_p p^{-1}) \prod_{p \notin S, p \leq x} (1 - a_p p^{-1} + p^{-1}).$$

Now let's look at N_p/p. For $p \in S$, this quotient is $(p - a_p)/p = 1 - a_p p^{-1}$. For $p \notin S$, it is $(p - a_p + 1)/p = 1 - a_p p^{-1} + p^{-1}$. Lo and behold! In every case, it is the corresponding factor in the product we just wrote down. So we can rewrite this product as

$$\frac{1}{\text{RHS up to } x} = \prod_{p \leq x} \frac{N_p}{p}.$$

We recognize the right-hand side now as $\beta_E(x)$. Taking reciprocals again we have

$$\text{RHS up to } x = \frac{1}{\beta_E(x)}.$$

Since we are only playing around, we can now let x tend to ∞. We better put back the scare quotes, for the same reason as before, but morally speaking we get

$$\text{``}L(E, 1)\text{''} \quad \text{``} = \text{''} \quad \text{``} \lim_{x \to \infty} \frac{1}{\beta_E(x)} \text{.''}$$

This "equation" says that how fast $L(E, s)$ tends to zero as $s \to 1$ might be related to how fast $\beta_E(x)$ tends to ∞ as $x \to \infty$. A hunch based along these lines led Birch and Swinnerton-Dyer to equate the analytic rank of E and the algebraic rank of E in their conjecture, the connecting link being the rate of growth of $\beta_E(x)$.

Amazingly, the American mathematician Dorian Goldfeld (1947–) proved a theorem, mentioned earlier, that justifies this hunch. He needed to assume that E was modular, but we now know that to be true. Here is Goldfeld's result (Goldfeld, 1982):

THEOREM 15.2: Let E be an elliptic curve defined over $\mathbf{Q}$. Suppose

$$\beta_E(x) \sim C(\log x)^\rho$$

for some constant C. Then ρ is the analytic rank of E.

In particular, if $\beta_E(x)$ remains bounded as x grows, then $\rho = 0$.

4. The BSD Conjecture

We can now put everything together and state the BSD Conjecture in its weak form. In order to insure that you know exactly what you need to do to win the million dollars, we will quote the statement from Wiles's official enunciation in the Clay Mathematics Institute website. (We alter Wiles's statement slightly because we denote the elliptic curve by E and the analytic rank by ρ, and we include the terms of the L-function for all finite primes.) The conjecture was originally made in Birch and Swinnerton-Dyer (1965).

CONJECTURE 15.3 (Birch–Swinnerton-Dyer): The Taylor expansion of $L(E, s)$ at $s = 1$ has the form

$$L(E, s) = c(s - 1)^\rho + \text{ higher order terms}$$

with $c \neq 0$ and $\rho = $ [the algebraic] rank of $E(\mathbf{Q})$.

In other words:

The weak Birch–Swinnerton-Dyer Conjecture says that for any elliptic curve E defined over Q, the algebraic rank r of E and the analytic rank ρ of E are equal.

As we said before, there is also a strong or refined form of the BSD Conjecture. In addition to the statement above, it includes a formula for the leading coefficient c of $L(E, s)$ at $s = 1$ in terms of various complicated properties of E. To explain this formula would require an intolerably long and probably incomprehensible annex to this book. To get the flavor of the strong form, we can quote Wiles again (with slight variants of notation), without attempting to explain further the terms he uses (except to explain that he writes $|S|$ where we write $\#S$ for the number of elements in a finite set S). The refined form of the BSD Conjecture predicts that [...]

$$c = |\text{III}_E| R_\infty w_\infty \prod_{p|2\Delta} w_p / |E(\mathbf{Q})_{\text{tors}}|^2. \tag{15.4}$$

Here $|\text{III}_E|$ is the order of the Tate–Shafarevich group of the elliptic curve E, a group that is not known in general to be finite although it is conjectured to be so. It counts the number of equivalence classes of homogeneous spaces of E which have points in all local fields. The term R_∞ is an $r \times r$ determinant whose matrix entries are given by a height pairing applied to a system of generators of $E(\mathbf{Q})/E(\mathbf{Q})_{\text{tors}}$. The w_p's are elementary local factors and w_∞ is a simple multiple of the real period of E.

There is a strong belief that the analytic rank of a "random" elliptic curve is equally likely to be odd or even. Therefore, if we believe the BSD Conjecture, the algebraic rank of a random elliptic curve is also equally likely to be odd or even. We used this remark in section 3 of chapter 10.

Much spectacular work has been performed trying to prove the BSD Conjecture. A certain amount of it can be summarized in the following theorem:

THEOREM 15.5: If the analytic rank of E is 0 or 1, then Conjecture 15.3 holds for E.

In particular, given an elliptic curve E, if $\beta_E(x)$ remains constant as x grows, then the algebraic rank is 0. Theorem 15.5 verifies this part of Birch and Swinnerton-Dyer's original insight. Of course, at this point, most number theorists expect the whole BSD Conjecture to be true.

Theorem 15.5 is the work of many people, and again we refer to Wiles (2006) for a short summary of the ingredients of this theorem.

If E is our example elliptic curve $y^2 = x^3 - x$, it is known that $L(E, 1) = 0.65551\ldots \neq 0$, and therefore the analytic rank is 0. Theorem 15.5 now assures us that $\#E(\mathbf{Q})$ is finite.

5. Computational Evidence for BSD

There is a variety of computational evidence for the BSD Conjecture. This goes back to Birch and Swinnerton-Dyer themselves. The computations are all for certain individual elliptic curves or particular types of elliptic curves.

The computations you can do, given a particular E:

1. You can compute $\beta_E(x)$ for x not too large, because it is easy to compute the N_p's by solving congruences modulo p. By graphing the logarithm of $\beta_E(x)$ versus the logarithm of the logarithm of x, and viewing the average slope of the graph, you can get a good guess for how fast $\beta_E(x)$ is growing, and therefore what the analytic rank ought to be.

2. You can use integration formulas coming from the theory of modular forms to compute $L(E, s)$ near $s = 1$ and see whether it looks as if $L(E, 1)$ is 0 or not. If it appears that $L(E, 1) = 0$, you can then compute $L(E, 1 + h)/h$ for small h and see if that ratio looks like 0. Assuming that $L(E, 1)$ really is zero, the Taylor series for $L(E, s)$ near $s = 1$ will look like

$$a_1(s - 1) + a_2(s - 1)^2 + \cdots .$$

Then

$$L(E, 1 + h)/h = a_1(1 + h - 1)/h + a_2(1 + h - 1)^2/h \cdots$$
$$= a_1 + a_2 h + \cdots .$$

So if $L(E, 1 + h)/h$ for small h is far from 0, you can guess that $a_1 \neq 0$, and therefore the analytic rank of E is exactly 1. On the other hand if $L(E, 1 + h)/h$ for small h is very close to 0, you can guess that $a_1 = 0$ also, and therefore the analytic rank of E is at least 2. In the latter case, you can play this game again, now looking at $L(E, 1 + h)/h^2$. And so on, until you find a power ρ such that $L(E, 1 + h)/h^\rho$ is far from 0 for very small h, whereas $L(E, 1 + h)/h^k$ is very close to 0 for very small h for all nonnegative integers k less than ρ. Under these circumstances, you have a very good guess that the analytic rank of E is exactly ρ. But these computations are not a proof that it is exactly ρ.

3. There are various upper and lower bounds proven for the algebraic rank. If you are lucky, for a particular E these might coincide, and then you will know its algebraic rank. For example, if you know some upper bound of 7 for the algebraic rank of E and then you actually find 7 independent points, you can conclude that the algebraic rank of E equals 7 on the nose. Then you probably can compute that the L-function of E looks like it vanishes to order 7 at $s = 1$, and thus give evidence that the BSD Conjecture is probably true for this E.

   Thus, in many cases, you can compute the algebraic rank of E. But usually you cannot *prove* that the analytic rank is the number that you estimated numerically, except when the analytic rank is small. (All known provable examples have analytic rank ≤ 3.)

4. In some cases, you can both compute the algebraic rank of a particular elliptic curve E and prove that the analytic rank equals the algebraic rank. Thus, you can verify the weak BSD Conjecture for that curve E. For example, the elliptic curve given by $y^2 = 4x^3 - 28x + 25$ can be proven to have both algebraic rank and analytic rank equal to 3.

Beginning in 1958, Birch and Swinnerton-Dyer used a computer and some clever theory to find the ranks of various elliptic curves. They expressed an interesting caveat in Birch and Swinnerton-Dyer (1963):

> …we believe that results obtained from a computer should not be automatically trusted. In some cases they can be checked because they have properties which were not essentially used in the course of the calculation and which would be unlikely to survive if an error had been made. (For example, if a table of a smooth function has been calculated without the use of interpolation, it can be checked

by differencing.) But if checks of this sort are not available, results should not be fully trusted until they have been independently reproduced by a different programmer using a different machine. We do not think this sets an unreasonable standard, now that computers are becoming so widely available; and we are satisfied that lower standards have already led to a number of untrue results being published and believed.[5]

They found that by computing $\beta_E(x)$ for various elliptic curves E and inspecting the rate of growth, they could often predict the algebraic rank of E "with fairly consistent success," as they put it in Birch and Swinnerton-Dyer (1965).

Starting in 1960, Birch and Swinnerton-Dyer were able to give evidence for their conjecture for certain elliptic curves, all of a special type called "having complex multiplication." For elliptic curves with complex multiplication, they found a rapidly convergent series that allowed them to get a good idea as to whether $L(E, s)$ was zero or not at $s = 1$. This is because for such curves, there is actually an explicit formula for the N_p's. As related in the introduction to Birch and Swinnerton-Dyer (1965), British mathematician Harold Davenport (1907–69) then showed them a better way to compute $L(E, s)$ at $s = 1$, and this improvement enabled them to make the stronger version of their conjecture.

In his official problem statement for the Clay Institute's Millennium Problems, Wiles, following Russian mathematician Yuri Manin (1937–), points out that the strong form of the BSD Conjecture can be used effectively to find points of infinite order on $E(\mathbf{Q})$. The weak form only tells you how many independent points to look for.

6. The Congruent Number Problem

We conclude by giving a concrete example of a number theoretical problem that can be solved if the BSD Conjecture is proven to be true. This was shown by the American mathematician Jerrold Tunnell (1950–).

---

[5] Computers were much more primitive in the 1950s and 1960s than they are now. But human fallibilities are the same, and this is probably still good advice. Birch and Swinnerton-Dyer did not give examples of untrue results they thought were being published and believed.

Here's the congruent number problem, which dates back at least to Arab mathematicians in the Middle Ages. We give you an integer n. Say $n = 1$ or $n = 5$ or $n = 1234567$. Can you find a right triangle with sides a, b, and c all *rational* numbers with area n? In other words, you need to solve the simultaneous equations

$$\begin{cases} a^2 + b^2 = c^2 \\ \dfrac{ab}{2} = n \end{cases}$$

for rational numbers a, b, and c. If you can find a solution, then n is called a "congruent number." It's not too hard to see that n is a congruent number if and only if the algebraic rank of the elliptic curve defined by the equation $y^2 = x^3 - n^2 x$ is greater than 0, that is, if that curve has infinitely many rational points on it. Each rational point leads to a different solution (a, b, c).

On pp. 110–111 of *Fearless Symmetry*, we showed how the congruent number problem leads to an elliptic curve when $n = 1$. The idea works for all n.

Set $X = \frac{a}{c}$, $Y = \frac{b}{c}$. The congruent number problem becomes

$$\begin{cases} X^2 + Y^2 = 1 \\ \dfrac{XY}{2} = \dfrac{n}{c^2}. \end{cases}$$

In the Prologue, we saw that except for $X = -1$, $Y = 0$, all solutions of the first equation are of the form

$$X = \frac{1 - t^2}{1 + t^2}$$

$$Y = \frac{2t}{1 + t^2}$$

Using these values and setting $w = 1 + t^2$, we just need to solve the second equation, which becomes

$$\frac{(1 - t^2)(2t)}{2w^2} = \frac{n}{c^2}$$

or equivalently

$$t - t^3 = n \left(\frac{w}{c} \right)^2.$$

Although w depends on t, c is arbitrary, so we can set $u = \frac{w}{c}$ and we need to solve

$$t - t^3 = nu^2.$$

This is a nonsingular cubic equation. If you want to put it in the standard form of an elliptic curve, set $t = -\frac{x}{n}$ and $u = \frac{y}{n^2}$ and the equation becomes $-n^{-1}x + n^{-3}x^3 = n^{-3}y^2$. Multiply through by n^3 to obtain

$$y^2 = x^3 - n^2 x.$$

Tunnell showed that the truth of the BSD Conjecture gives you an explicitly computable criterion for whether or not n is a congruent number. He gave the criterion, which you can find in Tunnell (1983); we follow the exposition in Conrad (2008). Besides containing an elegant exposition of the subject with lots of examples, this article explains clearly the origin of the term "congruent number." The connection between the congruent number problem for a positive integer n and the elliptic curve $y^2 = x^3 - n^2 x$ is given by the following elementary theorem.

THEOREM 15.6: Let n be a positive integer. There is a one-to-one correspondence between the two sets

$$\left\{ (a, b, c) \; \middle| \; a^2 + b^2 = c^2, \; \frac{ab}{2} = n, \; a, b, c > 0 \right\}$$

$$\Leftrightarrow \left\{ (x, y) \; \middle| \; y^2 = x^3 - n^2 x, \; x, y > 0 \right\}$$

given by the mutually inverse functions

$$(a, b, c) \mapsto \left(\frac{nb}{c - a}, \frac{2n^2}{c - a} \right) \qquad (x, y) \mapsto \left(\frac{x^2 - n^2}{y}, \frac{2nx}{y}, \frac{x^2 + n^2}{y} \right).$$

The verification of this theorem is a complicated exercise in elementary algebraic manipulations. Tunnell's Theorem is much more subtle and difficult to prove. Here is one way to phrase it.

THEOREM 15.7: Suppose that n is a squarefree positive integer. (That is, n is not divisible by any square other than 1.) If n is odd, compute n_1 and n_2, while if n is even, compute n_3 and n_4, applying the formulas

$$n_1 = \#\{(x, y, z) \mid n = x^2 + 2y^2 + 8z^2\}$$

$$n_2 = \#\{(x, y, z) \mid n = x^2 + 2y^2 + 32z^2\}$$

$$n_3 = \#\{(x, y, z) \mid \tfrac{n}{2} = x^2 + 4y^2 + 8z^2\}$$

$$n_4 = \#\{(x, y, z) \mid \tfrac{n}{2} = x^2 + 4y^2 + 32z^2\}$$

where in all 4 cases, the sums are over all integers x, y, and z.

If n is odd and a congruent number, then $n_1 = 2n_2$. If n is even and a congruent number, then $n_3 = 2n_4$. Moreover, if the weak form of the Birch–Swinnerton-Dyer Conjecture is true, then the converse is also true: If n is odd and $n_1 = 2n_2$, then n is a congruent number, and if n is even and $n_3 = 2n_4$, then n is a congruent number.

The proof is very far beyond what we can cover in this book. It uses deep properties of modular forms.

One interesting consequence of the connection to elliptic curves is that it implies the following otherwise very nonobvious fact: If n is a congruent number, there are infinitely many different right triangles with rational sides whose area is n. By the way, 1 is not a congruent number, but 5 is.

EXERCISE: Assume the BSD Conjecture. Is 1234567 a congruent number?

SOLUTION: Assume that n is odd and $n = x^2 + 2y^2 + 8z^2$. Then x must be odd and hence $x^2 \equiv 1 \pmod 8$. Also, y is either even or odd, and so $2y^2$ is congruent to either 0 or 2 $\pmod 8$. It follows that

n is congruent to either 1 or 3 (mod 8). The same reasoning shows that if n is odd and $n = x^2 + 2y^2 + 32z^2$, then n is again congruent to either 1 or 3 (mod 8).

Now, $1234567 = 127 \cdot 9721$, and both of those factors are prime, so we know that 1234567 is squarefree. Moreover, $1234567 \equiv 7$ (mod 8), telling us that $n_1 = 0$ and $n_2 = 0$. Therefore, we know that $n_1 = 2n_2$. Assuming the BSD Conjecture and using Tunnell's theorem, we conclude that 1234567 is indeed a congruent number.

Similar reasoning shows that (again assuming that the BSD Conjecture is true) if n is a squarefree integer, and n leaves a remainder of 5, 6, or 7 when divided by 8, then n is a congruent number.

EPILOGUE

. . . . .

Retrospect

We could have explained the Birch–Swinnerton-Dyer Conjecture in many fewer pages than we have used in this book. We wanted to move at a leisurely pace and give a bigger and more detailed picture. The BSD Conjecture has its natural context within the larger scope of modern algebraic geometry and number theory. Our story exemplifies how number theoretic problems that are very elementary to state can easily give rise to new and important mathematical ideas. Some of the methods we've explained were perhaps not obvious but after you get used them, they seem to be natural. As some philosopher said, habit is second nature.

Because the degree of a curve is such an important concept, we devoted many chapters to explaining that concept. We desired to regularize the behavior of a curve of a given degree with respect to how many points will be cut out on it by an arbitrary line. This desire led us to the concepts of *algebraic closure*, the *projective plane* (which is a kind of geometric closure), and *intersection multiplicity*. Each of these concepts is enormously useful in many areas of mathematics, in and beyond number theory.

Intersection multiplicity is a kind of sophisticated bookkeeping of how two curves intersect at a point. *Generating functions*, which bundle up various counts of solution sets, are another kind of bookkeeping we discussed. The *rank* of an abelian group, which enables us to distinguish algebraically different "sizes" of countably infinite groups, could be viewed as a third kind of bookkeeping.

In the meantime, we began to study an elliptic curve E. The full description was made possible by using our earlier work: algebraic closures, the projective plane, and intersection multiplicities. In this way, we could speak of the points of E over $\mathbf{C}$ and over the algebraic closure of $\mathbf{F}_p$, including the point at infinity, and we could give a uniform explanation of the group law on E.

As we said, we needn't have been so lavish. Without the complex numbers, we could have limped along with just the real numbers, because we were most interested in the rational points on E and rational numbers are all real. Without the projective plane, we could simply have stipulated a "point at infinity" on E. Without intersection multiplicity, we could have defined what it means to add a point on E to itself directly and we could have studied singular cubic curves by using more algebra. But not only does our broader exposition enhance the understanding of E itself, we also think that it was worth while to show how mathematicians are led to powerful developments of elementary objects, such as fields, planes, and intersections.

Given an elliptic curve E, we defined its L-function, $L(E, s)$, a generating function that keeps track of how many points there are on the curve over various finite fields. The set of rational points on the curve turned out to be an abelian group, so that its rank measures how many rational points there are on the curve. This is the *algebraic rank* of the curve. On the other hand, we defined its *analytic rank* as the order of vanishing of $L(E, s)$ at $s = 1$. At least heuristically, the analytic rank measures the number of points modulo p on the curve for various p. Therefore, we had reason to expect that if the algebraic rank is large, the analytic rank should also be large.

This expectation is refined in the Birch–Swinnerton-Dyer Conjecture, which states that the two ranks should be equal. So we finished up by explaining the BSD Conjecture, giving a little bit of its history, and a little bit of evidence for its correctness.

Let's bring this all down to earth with an example in the style of Diophantus. QUESTION: When can the difference of a rational number and its cube be a square? If possible, give all solutions.

We can formulate this problem algebraically as asking for rational solutions of the equation $y^2 = x^3 - x$. That equation defines an elliptic curve. We saw at the end of chapter 15 that we know the weak form of the BSD Conjecture to be true for this curve, and that the analytic rank, and therefore the algebraic rank, is 0. In other words, $E(\mathbf{Q})$ is a finite set, and using our knowledge of the group structure, that tells us that $E(\mathbf{Q})$ consists only of torsion points. Now the Nagell–Lutz Theorem 10.2 tells us that all torsion points must have integer coordinates, and moreover that either y is 0, or else that y divides Δ_E, which for this curve is -64. If $y = 0$, we can easily solve the equation $x^3 - x = 0$, and see that there are three x-values: 0, 1, and -1. Otherwise, y must be a power of 2 dividing 64, and so $y^2 \leq 4096$. We must have $x^3 \leq 4096$, and so $x \leq 16$. We can now substitute each of the values 2, 3, $\ldots$, 16, for x, and see that none of them give an integer value for y. Therefore, the only answers to our original question are $x = 0$, $x = 1$, and $x = -1$.

Where Do We Go from Here?

First, there is the proof of the BSD Conjecture itself. It has been proven to be true if the analytic rank of the elliptic curve is known to be 0 or 1. Many number theorists are nibbling along the edges of the general case, but it seems likely that a new idea or ideas will be necessary before the general case is proven. Perhaps major results will continue to be proven piecemeal, for example, for curves of analytic rank 2. Or maybe someone will prove the whole conjecture at one blow. Who knows if it will take 1, 10, 100, or 1,000 years?

Second, we can look at other polynomial equations in two variables with rational coefficients. Consider, for example, nonsingular equations with at least one rational solution. If the degree is 1, we have a line, and solving such an equation is the same as doing elementary arithmetic. The case of degree 2 we did in the Prologue. Curves of degree 3 are elliptic curves, and that's what this book was all about.

Now comes something of a surprise: A theorem of Faltings states that a nonsingular curve of degree 4 or higher can have only finitely many solutions. Does the set of solutions have any structure? Can any be found? Can all of them be exhaustively enumerated? It may be that no satisfying general scheme or set of algorithms will be discovered to answer these questions. However, L-functions can be defined for these curves also, and there is a generalization of the BSD Conjecture for them.

In fact, for any system of polynomial equations in any number of variables with rational coefficients, there are various profound generalizations of the BSD Conjecture, due to various mathematicians. Probably the most relevant of these is the Conjecture of Bloch and Kato, formulated by the American mathematician Spencer Bloch (1944–) and the Japanese mathematician Kazuya Kato (1952–). It was first published in 1990 and it implies the BSD Conjecture. Unfortunately, such topics are way beyond what we can explain in this book.

A funny thing about mathematics is how it goes on forever. No matter what we figure out there are always new things to be figured out. We are always at the edge of our abilities. We can clean up old areas of mathematics and discover new areas, and there keep arising yet newer methods, ideas, and problems. This vitality is a function of human character combined with the infinite inexhaustibility of the subject matter. Individual mathematicians have intense curiosity about mathematics. As a general rule, they cooperate freely with each other. However, many mathematicians are also intensely competitive. These two forces have always driven mathematics forward, and there is no sign they are abating.

We will never say: "There, we've proved that, now we know everything about mathematics." We will never know enough, let alone everything, about mathematics.

Bibliography

* * * * *

Ash, Avner, and Robert Gross, *Fearless Symmetry: Exposing the Hidden Patterns of Numbers*, Princeton University Press, NJ, 2006. With a foreword by Barry Mazur.

Birch, Bryan John, and Henry Peter Francis Swinnerton-Dyer. 1963. Notes on elliptic curves. I, *J. Reine Angew. Math.*, **212**, 7–25.

———1965. Notes on elliptic curves. II, *J. Reine Angew. Math.*, **218**, 79–108

Breuil, Christophe, Brian Conrad, Fred Diamond, and Richard Taylor. 2001 On the modularity of elliptic curves over **Q**: Wild 3-adic exercises, *J. Amer. Math. Soc.*, **14**, no. 4, 843–939.

Carlson, James, Arthur Jaffe, and Andrew Wiles (eds.), *The Millennium Prize Problems*, Clay Mathematics Institute, Cambridge, MA, 2006. Available at http://www.claymath.org/library/monographs/MPP.pdf.

Conrad, Keith. 2008. The congruent number problem, *Harv. Coll. Math. Rev.*, **2**, no. 2, 58–74. Available at http://thehcmr.org/node/17.

Courant, Richard, and Herbert Robbins, *What is Mathematics?: An Elementary Approach to Ideas and Methods*, Oxford University Press, New York, 1979.

Davenport, Harold, *The Higher Arithmetic: An Introduction to the Theory of Numbers*, 8th ed., Cambridge University Press, 2008. With editing and additional material by James H. Davenport.

Davis, Philip J. 1959. Leonhard Euler's integral: A historical profile of the gamma function, *Amer. Math. Monthly* **66**, 849–869.

Derbyshire, John, *Prime Obsession: Bernhard Riemann and the Greatest Unsolved Problem in Mathematics*, Plume, New York, 2004. Reprint of the 2003 original.

Devlin, Keith, *The Millennium Problems: The Seven Greatest Unsolved Mathematical Puzzles of Our Time*, Basic Books, New York, 2002.

Euclid, *The Thirteen Books of the Elements*, 2nd ed., translated by Thomas L. Heath, Dover, New York, 1956.

Goldfeld, Dorian. 1982. Sur les produits partiels eulériens attachés aux courbes elliptiques, *C. R. Acad. Sci. Paris Sér. I Math.* **294**, no. 14, 471–474 (French, with English summary).

Heath, Thomas L., *A History of Greek Mathematics, From Thales to Euclid*, Vol. 1, Dover, New York 1981.

Lakatos, Imre, *Proofs and Refutations: The Logic of Mathematical Discovery* (John Worrall and Elie Zahar, ed.), Cambridge University Press, 1976.

Livio, Mario, *The Golden Ratio: The Story of ϕ, the World's Most Astonishing Number*, Broadway Books, New York, 2002.

Rubin, Karl, and Alice Silverberg. 2002. Ranks of elliptic curves, *Bull. Amer. Math. Soc. (N.S.)*, **39**, no. 4, 455–474. Available at `http://www.ams.org/journals/bull/2002-39-04/S0273-0979-02-00952-7/`.

Silverman, Joseph H., *The Arithmetic of Elliptic Curves*, 2nd ed., Graduate Texts in Mathematics, Vol. 106, Springer, Dordrecht, 2009.

Silverman, Joseph H., and John Tate, *Rational Points on Elliptic Curves*, Undergraduate Texts in Mathematics, Springer-Verlag, New York, 1992.

Taylor, Richard, and Andrew Wiles. 1995. Ring-theoretic properties of certain Hecke algebras, *Ann. of Math. (2)*, **141**, no. 3, 553–572.

Thomas, Ivor (trans.), *Greek Mathematical Works*, revised, Vol. 1, Harvard University Press, Loeb Classical Library, Cambridge, MA, 1980.

Titchmarsh, Edward Charles, *The Theory of the Riemann Zeta-Function*, 2nd ed., The Clarendon Press, Oxford University Press, New York, 1986. Edited and with a preface by David Rodney Heath-Brown.

Tunnell, Jerrold B. 1983. A classical Diophantine problem and modular forms of weight 3/2, *Invent. Math.*, **72**, no. 2, 323–334.

Weil, André, *Number Theory: An Approach Through History from Hammurapi to Legendre*, Modern Birkhäuser Classics, Birkhäuser Boston Inc., MA, 2007. Reprint of the 1984 edition.

Wiles, Andrew. 1995. Modular elliptic curves and Fermat's last theorem, *Ann. of Math. (2)*, **141**, no. 3, 443–551.

——2006. The Birch and Swinnerton-Dyer conjecture, The Millennium Prize Problems, Clay Math. Inst., Cambridge, MA, pp. 31–41. Available at `http://www.claymath.org/library/monographs/MPP.pdf`.

Index